BULLETIN

OF THE

NATIONAL RESEARCH COUNCIL

Vol. 4, Part 7.   December, 1922   Number 25

CELESTIAL MECHANICS

A SURVEY OF THE STATUS OF THE DETERMINATION OF THE GENERAL PERTURBATIONS OF THE MINOR PLANETS.

*Appendix to the Report of the Committee on Celestial Mechanics, National Research Council**

BY A. O. LEUSCHNER

Professor of Astronomy, University of California

CONTENTS

*This Committee of the Division of Physical Sciences of the National Research Council consists of the following members: E. W. Brown, Professor of Mathematics, Yale University, Chairman; G. D. Birkhoff, Professor of Mathematics, Harvard University; A. O. Leuschner, Professor of Astronomy, University of California; H. N. Russell, Professor of Astronomy, Princeton University.

INTRODUCTION.

With approximately one thousand asteroids discovered and believed to be sufficiently observed to permit of fairly reliable orbit determinations, as indicated by the permanent numbers assigned to them, the task of preserving these discoveries has grown so stupendous that the time seems to have arrived for an analysis of the present astronomical practice in providing the necessary additional observations and calculations.

Hitherto, the burden of correcting orbit elements and computing ephemerides has rested principally on the Berlin Recheninstitut. In recent years the Marseilles Observatory has rendered notable service in contributing orbits and ephemerides. Observations, photographic and visual, are regularly made at a number of observatories. The Berlin Recheninstitut publishes ephemerides and other results in the Astronomische Nachrichten, and in the Ephemeriden der Kleinen Planeten. Up to 1918 these data appeared also in the Astronomisches Jahrbuch. The number of oppositions during which the minor planets have been observed, and the status of orbit determinations are annually summarized in the Vierteljahrschrift der Astronomischen Gesellschaft. In 1901 Bauschinger published the latest reliable elements, etc., with data concerning the perturbations for the then known 463 planets, "Tabellen zur Geschichte und Statistik der Kleinen Planeten."

The latest available collection of elements is contained in the Berlin Jahrbuch for 1918. The adopted Jahrbuch elements serve the purpose of providing ephemerides from opposition to opposition. Their origin may be traced from the notes given from year to year in the Jahrbuch, ending with 1918, and in *Kleine Planeten*. Some of the elements include arbitrary corrections to the mean motion and to the mean anomaly for the purpose of representing late oppositions so as to serve for prediction of immediately following oppositions. In other cases, approximate or accurate perturbations are included, with or without correction of the elements by the usual least squares adjustment. For thirty-six planets the elements in the Jahrbuch of 1918 are mean or osculating elements, derived in connection with general perturbations which are approximately included in the prediction of ephemerides. Until similar fundamental data shall have become available for the remaining planets, the present practice of the Recheninstitut appears to furnish the only certain method for the preservation of planetary discoveries.

In addition to the general pertubations of the thirty-six planets which are being used by the Jahrbuch, the general pertubations of a number of other planets have been derived on the basis of what, at the time, appeared to be reliable osculating elements. These presumably valuable data have been replaced by later elements derived independently, more or less accurately, with or without general perturbations, or upon the basis of arbitrary corrections. Bauschinger's Tabellen form a valuable key to some of these investigations, but even in the Tabellen the elements and perturbations cited do not, in all cases, represent the best elements and perturbations available, although perhaps in every case they are the most reliable for subsequent oppositions. This arises from the fact that earlier investigations were abandoned by Bauschinger in favor of later ones. Preliminary calculations have shown, however, that some of the earlier elements and perturbations represent distant oppositions at a later date more satisfactorily than his adopted elements and perturbations represent earlier oppositions equally remote.

Of great importance for the program of the Recheninstitut are the contributions of Brendel, who has developed methods for the approximate determination of the perturbations for certain groups of planets. Perturbations greater than 3'.4 within fifty years are included, with the object of reproducing geocentric places within 20' for 100 years. So far the necessary data have been published by Brendel, Labitzke, and Boda for 230 planets, approximately 25 per cent of the total number of known minor planets. The advantage to be gained from Brendel's contributions for these planets is that for the practical purpose of preserving these planets, by following their motion, it should become unnecessary as a rule to compute special perturbations for them, or even to apply corrections to the elements. Brendel plans to continue the work of supplying instantaneous elements and approximate perturbations for other groups of planets so that the program of the Recheninstitut, of the Marseilles Observatory, and of various investigators who, from time to time, publish improved elements and perturbations for ephemeris purposes, will become more and more simplified.

The preservation of planetary discoveries by observation and prediction with the aid of approximate perturbations is not the ultimate aim of astronomical science, but a necessary and unavoidable means to the end. The ultimate aim rests on the determination of mean elements and general perturbations which hold for all time or at least for very long periods within the limits of accuracy set by observation. It is expected that the elements and perturbations determined under the Newtonian law of gravitation may serve this purpose, provided that the mathematical difficulties will not prove insurmountable. It may be assumed that the rigid mathematical methods hitherto developed are satisfactory for planets with moderate eccentricity and inclination which are not in a very near commensurable ratio with any of the major planets, but it has not been established so far whether an accurate application of the Newtonian law would fully account for the motion of the minor planets even in the ordinary cases just referred to.

Exhaustive researches are available only for a very limited number of planets. Among these are (4) Vesta, (13) Egeria and (447) Valentine. The researches on (4) Vesta are due to Leveau, whose extraordinary investigations extend approximately over a complete century of oppositions. In connection with his work on the motion of Vesta, Leveau has aimed at a determination of the masses of Jupiter and Mars. His final value is larger than the best available mass of Jupiter by approximately one one-thousandth. On account of the moderate perturbations, the motion of Vesta does not lend itself as well to a determination of the mass of Jupiter as the motion of other minor planets with very large perturbations. Any slight departure from the true mass of Jupiter, et cetera, can reveal itself through the motion of Vesta only in long intervals of time, which accounts for Leveau's gradual improvement of his adopted mass by successively including longer periods of observation. For the present his results may be considered fundamental and final, so far as this planet is concerned. No other case has been studied so exhaustively. Later predictions are well within the errors of observation, and not the slightest departure from the Newtonian law is noticeable. It remains, however, to establish the same result for planets with large perturbations, particularly for such planets as have a mean motion commensurable with that of Jupiter. To avoid the necessity of gradually improving the Jupiter mass by means of subsequent observations of Vesta, it appears advisable to base further predictions on the best determined values of the masses of the major planets. Vesta also furnishes an example of the weight to be assigned to observations in the early part of the last century.

Leveau's investigations furnish a striking example of the fundamental researches necessary for the promotion of astronomical science as distinguished from the generally accepted program of observation and prediction for the preservation of discoveries.

For the study and interpretation of planetary statistics, particularly with reference to the origin of minor planets, the explanation of the gaps, the question of stability and ultimate destiny, or in general regarding their place in any hypothesis concerning the solar system, final mean elements derived on the basis of accurate developments of the perturbations are most essential. Fragments of fundamental investigations of perturbations are available for a number of minor planets. The value of some of these has been vitiated by corrections made in connection with the accepted program of approximate prediction, such as, for example, the correction of an accurate set of osculating elements derived by special or general perturbations, to represent later oppositions, either without perturbations or by taking account only of approximate or incomplete perturbations.

Fundamental investigations here are understood to include the determination of osculating or mean elements from a limited number of oppositions with complete regard of the perturbations, either special or general, in so far as they may have been appreciable. In connection with the study of the data existing for a limited number of selected planets, it has been found that the failure of such elements and perturbations to represent future oppositions in some cases can be accounted for by the fact that the masses of the major planets were known at the time with insufficient accuracy. The mere correction of the perturbations therefore, for the latest known values of the masses may render such elements and perturbations far more satisfactory than they appeared to be at the time when they were discarded in favor of new determinations of elements with or without perturbations.

Freed from effects of changes which affect disadvantageously their permanent value the fragments of fundamental investigations referred to are of great importance as a basis for researches and their intelligent application will involve a vast saving in computational and theoretical work.

At present it appears next to hopeless to the investigator to adopt a profitable form of attack in connection with any of the older minor planets without an enormous expenditure of time in searching astronomical records. This accounts for the many duplications of effort and for the disregard of previous valuable investigations. If systematically undertaken, the task of bringing to light the important data available for a final determination of the elements and general perturbations of the minor planets, does not appear insurmountable. Once available, such research surveys will be invaluable and should prove an encouragement to research, particularly to young investigators.

The research surveys of the few planets which are given below are intended to serve as illustrations of the data which should be made easily accessible. No claim is made for the absolute completeness of these data. The time for active work, with the aid of a few assistants, to prepare these preliminary surveys has extended only over a little more than a month. A great mass of material had to be consulted which was found to be of no importance to the purpose in hand.. This is being preserved on cards for easy reference, if required at any time. Thus care has been taken to eliminate elements which would not be considered as fairly accurate osculating elements particularly those which have resulted from corrections on the basis of subsequent oppositions purely for ephemeris purposes, without complete consideration of the perturbations and of the earlier oppositions in the final adjustment. This policy, however, has not been adhered to strictly, partly for historical and theoretical reasons with reference to preliminary elements, and partly for other reasons with reference to later elements.

Whenever possible, the reasons for the abandonment of previous investigations are given, but in many cases no reasons could be found, at least not in the astronomical records available in the library of the University of California. Some of these reasons are probably to be found in the records available in the library of the Lick Observatory, but in the limited time it has not been possible to consult these or other additional records for this preliminary survey. An immense amount of fundamental work has been accomplished by the Berlin Recheninstitut, particularly in computing special perturbations and deriving osculating elements, but has been published only in part. The remainder reposes in the archives of the Recheninstitut. It may be assumed that the immense task of providing ephemerides has interfered with the publication of the accumulated material. Without this material, research surveys such as those presented here are not complete.

A simple way of accomplishing the introduction of the improved mass of disturbing planets referred to above, is to multiply the final sum of all the terms for each component of the perturbations by the ratio of the new to the old mass. Aside from the improvement which it may be possible to make to some of the older fundamental data, particularly those which are no longer used for ephemeris purposes, by the introduction of the best determined values of the masses of the major planets, it is probably possible to enhance their values still further by correcting the elements on the basis of the existing developments of the general perturbations, with the aid of subsequent oppositions and in case of appreciable changes in the elements, by also correcting the numerical coefficients in the general developments by differential methods.

After revision of independently determined elements and perturbations for separate series of fairly consecutive oppositions, disconnected by a gap including a number of oppositions to which neither series was extended either backward or forward, the separate fundamental investigations will, in some cases, probably be found to be entirely consistent and thus become of permanent value, such as Leveau's investigations on (4) Vesta, without involving extensive theoretical and numerical work. In other similar cases the correction of the elements and perturbations pertaining to fundamental investigations for groups of oppositions separated by considerable gaps, so as to represent the osculating data at a subsequent epoch, may establish satisfactorily the connection between one or more groups of oppositions for which elements and perturbations have been independently determined with accuracy. The mode of attack will vary with the available data for different planets, as indicated by the research surveys, which this discussion advocates. The resurrection of the classical contributions of the pioneer investigators of planetary perturbations on a permanent basis, should produce material of great value for the ultimate aims of astronomical science concerning planetary investigations.

The proposed program of fundamental investigations cannot supersede the present astronomical practice in caring for the minor planets in the immediate future, but as stated above it will be of great assistance for the practical purposes of prediction, and should gradually solve the now stupendous task of preserving planetary discoveries, while furnishing at the same time the data for the more fundamental aims of astronomical science.

For the majority of the minor planets, probably the application of four successive steps or processes will be necessary to preserve the discoveries until final elements and perturbations can be made available. The first step or process represents the present practice principally conducted by the Berlin Recheninstitut. The second step is illustrated by Brendel's plan of supplying instantaneous elements and approximate perturbations. The third step corresponds to the determination of the elements and perturbations of the Watson asteroids undertaken by Leuschner, which are intended to provide fairly accurate but not final results. Hansen's and the Bohlin-v. Zeipel methods have been found most practical and accurate in this connection. The fourth and final step is demonstrated by the fundamental work of Leveau on (4) Vesta. It is the object of this discussion to encourage researches similar to Leveau's, and by supplying samples of research surveys for a limited number of planets to pave the way for a comprehensive international program in this connection.

It cannot be too strongly emphasized that accurate osculating elements are absolutely essential for fundamental investigations of the perturbations. While this requirement is fully recognized, the prevailing practice of changing elements for immediate ephemeris purposes is apt to lead to erroneous interpretation of available elements. Mean elements, in general, can be determined only after osculating elements and perturbations shall have become available. Some investigators have adopted as approximate mean elements the average of elements published for more or less extensive series of oppositions, assuming that these elements represent fairly reliable osculating elements. Even if this were the case, it hardly ever occurs that a sufficiently large number of elements, uniformly distributed over the orbit, are available to guarantee that, in taking the average, the effect of the periodic terms is entirely eliminated. But, as previously stated, many of the apparently reliable sets of elements are not osculating, but inferior elements produced by arbitrary changes or with incomplete perturbations.

Practically the only reliable method of arriving at accurate initial osculating elements consists in representing the observations of a limited number of oppositions by taking into account the special perturbations and in testing the validity of the resulting elements for one or more oppositions following. Osculating elements thus obtained will rarely require later changes which would affect the coefficients of the general perturbations. No correction of such elements should be attempted, except on the basis of the determination of complete special or general perturbations. As it was not considered necessary, at the time, to adhere strictly to the foregoing principle in Leuschner's program for the determination of the perturbations of Watson's asteroids, allowances for slight inaccuracies may later become necessary for some of the Watson planets. In particular, corrections to the larger coefficients of the perturbations may be necessary for the planets for which the initial adopted elements, considered at the time as sufficiently accurate, were neither accurate mean elements nor accurate osculating elements.

Attention has recently been called, in the Proceedings of the National Academy of Sciences of 1921, Vol. 8, No. 7, p. 170, and in the report of the Committee on Celestial Mechanics of the National Research Council, Bulletin of the National Research Council, Vol. 3, Part 4, No. 19, June, 1922, to the extremely satisfactory results obtained for the planets (10) Hygiea, and (175) Andromache, by the application of Leuschner's revision of von Zeipel's tables for the Hecuba group. Further reference to the great importance of "Gruppenweise Berechnung der Stoerungen," inaugurated by Bohlin, may therefore be omitted here. The methods of Bohlin and his followers serve admirably in connection with the third of the four stages outlined above for the determination of fundamental results. In certain cases of limited eccentricity and inclination, they will, no doubt, lead to final results.

No claim is made that the planets for which research surveys are given below are the ones most in need of immediate attention. Further study of available data will be necessary to classify the planets with reference to the requirements of observation and computation, as outlined in the report of the American Committee on Comets and Asteroids, presented at the Brussels meeting of the International Astronomical Union in 1919; nor are the planets considered below the most important for fundamental scientific purposes. The list, however, may be considered as fairly representative of the immediate research requirements. To some extent the selection has been accidental. Thus the computing section of the British Astronomical Society has undertaken the computation of the ephemerides of the first four planets. In this connection it sought advice regarding the best available data and methods of procedure. The research surveys of the first four planets were undertaken to aid the computing section in its undertaking. The importance of the Trojan group is too well known to be emphasized. For further investigations concerning the theories of the six planets belonging to this group the research surveys given will be of considerable value. It is of interest to note that Leuschner's orbit methods as applied by Einarsson, appear to be the most promising for the determination of preliminary osculating elements, while Wilkens' method deserves careful trial in deriving the perturbations. E. W. Brown's unpublished theory promises to be thoroughly fundamental. For the two planets of the Trojan group last discovered, more accurate preliminary osculating elements are immediately needed. For other planets, the list of research surveys themselves will reveal the most necessary work to be done. In general, reference to theoretical investigations is included only in connection with a simultaneous new determination of elements. Thus the numerous and important investigations on the theory of the Trojan group are not considered here, the chief object of the surveys of these planets being to furnish numerical data and encourage their improvement as a basis for such theories.

The form in which the research surveys are presented must be considered experimental. That adopted is the outcome of several other attempts at presenting the material. It is hoped that this report will call forth helpful criticisms and suggestions which may ultimately lead to the adoption of some definite plan for international cooperation. Much material has been collected on planets not included in the list, which, it is hoped, may be printed later.

It was found that the research surveys for the various planets could not be made so complete that the investigator may abstain from referring to the sources themselves. This applies also to the collection of elements. The elements are collected merely for purposes of comparison and are not reproduced with uniform accuracy.

For practically all the planets in the list, except the first four and several others, a fairly complete bibliography of observations has been prepared, but this bibliography is published here only for the last two of the Trojan group.

Attention might well be called here to the need of curtailing indiscriminate observations. Even in recent years observations have been multiplied for planets for which two or three accurate observations at each opposition would be sufficient for all scientific purposes. It is planned to formulate in the near future definite proposals for an international program of observations.

The main purpose of this report is the encouragement of fundamental researches essential to the ultimate aims of astronomical science, which, for their consummation, require the knowledge of accurate elements and perturbations of the minor planets.

The surveys have been prepared in the main by Dr. W. F. Meyer, and by Dr. H. Thiele, assisted by several advanced students in astronomy, who have gathered the necessary references. For the Trojan group, unpublished data collected by Dr. Sturla Einarsson have been available.

As a rule the abbreviations adopted for the references are those of the Astronomischer Jahresbericht.

The usual notations of the elements are adhered to, both μ and n being used for the mean daily motion.

(1) CERES.

The first and largest of the minor planets was discovered 1801, January 1, by Piazzi in Palermo.[1]

Piazzi assumed that the object was a comet, but several astronomers succeeded in proving from the 22 meridian observations near the stationary point over an heliocentric arc of 9° that it was a planet moving in a nearly circular orbit; thus Burckhardt[2] computed Elements A, Olbers[3] the circular Elements B, Piazzi[4] the circular Elements C. Only the computation by Gauss,[5] Elements D, was accurate enough, especially in the determination of perihelion and eccentricity, to indicate where the planet might be found the following year.

Olbers found Ceres again 1802, January 1, ½° from the predicted place, near the place where, three months later, he discovered the second of the minor planets. The new observations naturally increased the accuracy of the elements notably; thus Gauss[6] computed Elements E, from observations in 1801, and January 1802; representation in February 1802, $+7''$ in a, $-20''$ in δ. Burckhardt[7] including the perturbations larger than 30' found Elements F.

For some years the orbit of Ceres was investigated by Oriani, Burckhardt, and Gauss by taking the perturbations into account, but the efforts of Gauss went farther than those of the others. Burckhardt[8] started with the computation of perturbations at intervals of two days, and later computed tables founded upon them. Oriani[9] used Laplace's method, with which also Gauss started. Gauss developed the perturbations first in 1802, together with Elements VIII, G, and formed tables of perturbations[10] and later in 1805[11] when he used the same interpolatory development of the perturbative function as Hansen later used in 1830.

The orbit computation was taken up later by Heiligenstein.[12] He derived Elements H from the oppositions 1818, 1820, 1821, 1822, 1825, 1826, 1827, with special perturbations of the elements by Jupiter, (mass 1/1053.924). Representation of the normal places $-10''$ to $+6''$ in mean longitude. Correction to the ephemeris for 1830 April, May, $-6''$ in a, $-10''$ in δ.

Heiligenstein's ephemeris deviates 15' from the ephemeris in B. J. 1830, which was based on the elements of Gauss (XIII, 1809), using the tables of perturbations by Gauss and an empirical correction by Encke of 14' to the mean longitude determined from the last observations.[13]

In B. J. for 1831 Encke[14] gives an ephemeris from new Elements I of his own based on the oppositions 1820, 1821, 1822, 1825. Jupiter mass 1/1053.924. Special perturbations by Jupiter only. Representation:

	1820	1821	1822	1825	1827	1829
in a	—6″	+2″	—4″	—3″	—2″	—27″
in δ	0″	0″	+6″	+1″	0″	11″

In B. J. 1832 to 1836 the ephemerides by Heiligenstein were published. Later the computation by Encke and Wolfers was used to 1871.

In the meantime Damoiseau[15] had given expressions for the perturbations containing a large number of terms but the individual coefficients do not seem to be very exact, according to Hill.

For the use of the American Ephemeris, E. Schubert[16] undertook to correct the elements by 250 observations in 14 oppositions, 1832-1854, using special perturbations of the elements by Jupiter as computed by Encke and Wolfers but corrected for the secular variation of the obliquity.[17] Elements J. Residuals in a —22″ to +21″, in δ —8″ to +8″; ϕ corrected according to A. J., Vol. 5, p. 73. A further correction of the elements by Schubert[18] was based on only four normal places in 1853, 1854, 1855, 1857; he applied the special perturbations of the elements by Jupiter and Saturn. Representation of the normals ±0″, "by which the correctness of the whole is proved." Elements K.

Godward[19] repeats the process of Heiligenstein, Encke, Wolfers, Schubert. The errors for fifteen oppositions 1857 to 1876 of the ephemerides in Nautical Almanac which include the perturbations of Venus, the Earth, Mars, Jupiter, Saturn gave by a least squares solution the Elements L. Ephemerides by these elements were given in the Nautical Almanac to 1913.

The corrections to Encke's ephemerides increased after 28 years to ±3ˢ in a ±20″ in δ.

The corrections to Schubert's ephemerides increased after 23 years to +6ˢ in a ±40″ in δ.

The corrections to Godward's ephemerides increased after 36 years to +2ˢ in a ±10″ in δ.

For the purpose of illustrating his modified form of computing absolute perturbations Hill[20] computed the first order perturbations of Ceres by Jupiter starting with the first elements by Schubert (uncorrected). It was found that the osculating mean motion differed widely from the mean mean motion. An arbitrary value was substituted. The Jupiter mass is taken to be 1/1047.355. The expressions for the periodic terms of the perturbations are given. In order to arrive at mean elements as well as to see how closely the perturbations represent the observations, ten normal places—1802, 1807, 1830, 1857, 1863, 1866, 1873, 1883, 1885, 1890, were formed. Secular perturbations of Mars, Jupiter, and Saturn were computed by the method of Gauss. The periodic perturbations by Mars and Saturn were taken from the tables of Damoiseau. Preliminary elements and a least squares solution led to mean Elements M. The residuals are —40″ to +40″ in hel. longitude, —20″ to +13″ in geoc. latitude. Hill originally intended to enlarge and complete his theory of Ceres; for this purpose he collected the observations into 75 normals from 1801-1897.[21] He published the positions because he did not expect to finish the work. The collection is not complete.

As an extension of Hill's work Merfield[22] has given a computation of the secular perturbations of Ceres arising from the action of the eight Major Planets. From Hill's theory and his mean elements using the method of Gauss as set forth by Hill the numerical values of the action of the planets were derived.

M. Wolf[23] has developed the expression (ρ) according to the theory of Gyldén in the case of Ceres. Cf. Tisserand, Mécanique Céleste, Vol. 4.

M. Viljev[24] has published tables of absolute perturbations of Ceres after the method of Hansen.

REFERENCES

[1] Bode's B. J. 1804, p. 249. B. J. 1805, p. 202. v. Zach, Monatliche Correspondance. Not available here. Lalande, Connaissance des Temps de l'annee xiii. Witt, H.u. E. vol. 14.

[2] Bode's B. J. 1804, p. 255.

[3] Bode's B. J. 1804, p. 256.

[4] Bode's B. J. 1804, p. 259.

[5] Bode's B. J. 1805, p. 94. Gauss. Werke Bd. 6, p. 200.

[6] Gauss. Werke Bd. 6, p. 207.

[7] Bode's B. J. 1805, p. 96.

[8] A. J. vol. 16, p. 57.

[9] v. Zach. Monatliche Correspondance 1802, Dec. Not available here.

[10] Gauss. Werke Bd. 7, p. 375.

[11] Gauss. Werke Bd. 7, p. 401.

[12] A. N. vol. 7, p. 413.

[13] B. J. 1830, p. 245.

[14] B. J. 1831, p. 277. A. N. vol. 27, p.177.

[15] Connaissance des Temps 1846. Additions p. 32. Not available here.

[16] A. J. vol. 3, p. 153; p. 162.

[17] Nautical Almanac 1837. B. J. 1838, p.286.

[18] A. J. vol. 5, p. 73.

[19] M. N. vol. 38, p. 119.

[20] A. J. vol. 16, p. 57.

[21] A. J. vol. 21, p. 51.

[22] M. N. vol. 67, p. 551.

[23] Wolf. Sur les termes élémentaires dans l'expression du rayonvecteur. Stockholm, 1890.

[24] Publications de l'Observatoire Central Nicolas, Poulkovo. Not available here.

TABLE 1.—*Elements—(1) Ceres*

Letter	Date	MT	L	π	Ω	i
			° ′ ″	° ′ ″	° ′ ″	° ′ ″
A....	1801 Jan. 1.3328	....	68 59 37	248 59 37	80 58 30	10 47
B....	1801 Jan. 1	....	68 35 51.5	....	80 22 45	11 3 36
C....	1801	....	68 46 41	....	80 46 48	10 51 12
D....	1801	Palermo....	76 28 14	150 33 20	81 2 35	10 36 30
E....	1801	Palermo....	77 27 31	145 57 15	80 58 40	10 37 57
F....	1802	....	155 32 35	146 44 37	81 5 35	10 36 52
G....	1801	Seeberg....	77 19 34.9	146 33 37	80 54 59	10 37 56.0
H....	1818 Oct. 15.0..	Göttingen..	28 21 52.291	148 2 14.084	80 48 32.192	10 38 21.682
I....	1822 Jan. 22.0..	Göttingen..	127 36 44.2	147 36 57.6	80 41 55.0	10 38 7.7
			M ° ′ ″			
J....	1854 Jan. 0	Washington.	113 22 25.08	148 55 23.41	80 50 50.79	10 37 8.54
K....	1854 Jan. 0	Washington.	113 18 22.40	148 56 3.72	80 50 31.11	10 37 4.76
L....	1854 Jan. 0	Washington.	113 22 11.73	148 55 26.54	80 50 31.06	10 37 5.81
			L ° ′ ″			
M....	1850 Jan. 0.0...	Greenwich..	309 30 32.4	148 28 32.5	80 48 5.6	10 37 6.2

Letter	Date	MT	φ	μ	Equinox	Author
			° ′ ″	′		
A....	1801 Jan. 1.3328	....	2 5	859.05	....	Burckhardt I
B....	1801 Jan. 1	....	0	786.528	....	Olbers
C....	1801	....	0	795.937	....	Piazzi
D....	1801	Palermo....	4 2 45	784.254	....	Gauss I
E....	1801	Palermo....	4 40 10	769.7925	....	Gauss VII
F....	1802	....	4 31 25	771.363	....	Burckhardt II
G....	1801	Seeberg....	4 31 17.8	770.7951		Gauss VIII
H....	1818 Oct. 15.0..	Göttingen..	4 31 5.183	771.2273825	1818.00	Heiligenstein
I....	1822 Jan. 22.0..	Göttingen..	4 31 18.0	770.72468	1810.00	Encke
J....	1854 Jan. 0	Washington.	4 24 28.41	769.63875		Schubert
K....	1854 Jan. 0	Washington.	4 24 29.36	769.62476	1854.00	Schubert
L....	1854 Jan. 0	Washington.	4 24 29.65	769.64746	1854.00	Godward
M....	1850 Jan. 0.0...	Greenwich..	4 29 57.8	770.718276	1850.00	Hill*

*Mean elements.

(2) PALLAS

Discovered by Olbers at Bremen 1802, March 28.[1] Olbers attempted to compute a circular and a parabolic orbit for the new planet, both of which failed. His computation showed the orbit had a large inclination and considerable eccentricity.

From observations extending from April 1 to July 8, Gauss[2] computed Elements A (Gauss V). They are improvements on preliminary sets. With these elements an ephemeris for 1803 was computed.

From observations extending from April 4 to May 20, Burkhardt[3] computed Elements B. With these elements Burkhardt computed the perturbations in longitude, latitude, and radius vector covering the period April 4 to May 20.

The planet was reobserved by Harding 1803, Feb. 21st. The comparison between Gauss' ephemeris and observations was as follows:

1803	Δa	$\Delta \delta$
Feb. 21	$+2'$ $02''$	$-34''$
Feb. 23	$+2$ 35	-57

On the basis of these residuals, Gauss[4] improved Elements A Gauss (V). These new Elements C (Gauss VI) represent the observations as follows:

1803	Δa	$\Delta \delta$
Feb. 21	$- 20''.0$	$+ 15''.8$
Feb. 23	$+ 7.8$	$- 7.7$

From a set of elements, based on oppositions 1804, 1805, 1807, 1808, Gauss[5] derived an improved set of Elements D from a least squares solution. This solution includes also the oppositions 1803 and 1809 and forms the basis for the computation of perturbations as outlined below.

Gauss[6] first attempted to construct tables of perturbations for the four known minor planets, but the large eccentricity and inclination forced him to formulate a theory for Pallas based on the variation of the elements expressed analytically and integrated by mechanical integration. From two successive calculations of the special perturbations, due to Jupiter, Gauss derived the improved Elements E, which represented the heliocentric longitudes for the first seven oppositions within $\pm 8''$.

In 1811 Gauss[6] began his first computation of general perturbations due to Jupiter. For this purpose he used Laplace's elements of Jupiter, epoch 1805, and his own elements of Pallas for the same epoch. This computation led to a set of mean Elements (F). With these mean elements for epoch 1810 and similar elements for Jupiter (Laplace) a second computation of general perturbations due to Jupiter was undertaken. This computation led to the following results:

The mean motion of Pallas oscillates between 18/7 of ♃ motion $\pm 0''.2153$, and 1894 revolutions of Pallas = 737 of Jupiter. A new value for Jupiter's mass = 1/1042.86. Then follows (1816-1817) the computation of perturbation tables due to Jupiter, Saturn and Mars. In this latter work, Gauss was assisted by Encke and Nicolai.

In Astronomisches Jahrbuch 1816, page 234, Bode gives the best set of elements by Gauss up to that time (Elements G).

About 1824, Encke[7] used Gauss' elements based on early oppositions and computed the perturbations due to Jupiter. He reports that Gauss' elements with Jupiter's perturbations represent the opposition of 1823 as follows:

1823	Δa	$\Delta \delta$
Oct. 9	$+13\overset{s}{.}2$	$+25\overset{''}{.}6$

He then gives Elements (H) for the epoch 1826, and with these computes the next ephemeris.

For the opposition in 1825, Encke[8] reports that the correction to the ephemeris is very large. But if the perturbations are included, the difference between observation and computation is as follows:

1825	Δa	$\Delta \delta$
March 23	$+42''.6$	$-33''.2$

He then gives a set of elements for the epoch 1827, and computes an ephemeris for 1827.

By 1834 Encke[9] reports a deviation of Pallas from computed places amounting to 5'. He states this may be due to use of Laplace's value for Jupiter's mass. It will be necessary to recompute elements covering all observations.

In A. N. No. 636, Encke publishes osculating Elements I for each year from 1831 to 1838; his fundamental starting elements are for the epoch of 1810, January 0. In B. J. 1838, p. 286, Encke draws attention to an error which he had committed in neglecting the corrections for the secular variation of the obliquity. In the British Nautical Almanac for 1837 Airy points out this error to which Encke refers.

Galle[10] undertook the reinvestigation of the orbit based on oppositions 1816, 1821, 1827, 1830, 1834, 1836, making use of Airy's value for the mass of Jupiter 1/1048.69. The former perturbations were retained except for the change due to Jupiter's mass. The resulting Elements J represent the heliocentric longitude and latitude as follows:

	1816	1821	1827	1830	1834	1836
ΔL	—19″	+34″	+4″	—14″	+5″	—13″
ΔB	— 1	+ 4	+6	—10	+1	+ 4

Galle states these differences may be accounted for if the perturbations of Saturn and Mars were taken into account.

In A. N. No. 636 osculating Elements K are published for each year from 1839 to 1850. These were computed by Galle. The starting elements are those for epoch 1810, January 0. In computing the special perturbations, Encke and Galle used mass of Jupiter 1/1053.924.

From 1851 to 1870 Galle[11] continues the special perturbations by Jupiter and later with the elements of Günther also those by Saturn. These were used in computing the ephemerides published in the Astronomiches Jahrbuch from 1862 to 1870. (See Elements L.)

Beginning with the year 1871 and continuing to 1919, the Jahrbuch published and used Farley's[12] osculating elements for computing the ephemeris. (See Elements M and N). Farley's computation includes the perturbations by Venus, Earth, Mars, Jupiter and Saturn. His computations are also the basis for the ephemerides published in the British Nautical Almanac. With Farley's elements we have the following comparisons:

Corrections to Ephemerides.

	1883	1892	1895	1906	1908	1914
Δa	—1ˢ.4	—1ˢ.2	—1ˢ.0	—2ˢ.5	—5ˢ.4	—2ˢ.1
$\Delta \delta$	+2″.7	+0″.7	+0″.8	+8″.2	—14.0	+4″.3

In Annales de l'Observatoire de Paris, Vol. I, Le Verrier publishes the results of his investigation on "Développement de la fonction perturbatrice relative à l'action de Jupiter sur Pallas. Calcul du terme dont dépend une inégalité à longue période du mouvement de cette dernière planète." Le Verrier states that the aphelion of Pallas is 54° from the intersection of the orbit with Jupiter. Consequently when Pallas is at aphelion the distance from Jupiter is increased on account of the great inclination of the orbit. This large inclination diminishes the effect due to the large eccentricity. Le Verrier gives the series for the reciprocal of the distance in a more convergent form and develops the equation in longitude depending on the argument 18$\mathcal{Q}$ — 7 Pallas. The maximum of the term is 895″. In his report before the Paris Academy,[18] Cauchy compares his theory with the results by Le Verrier; his value for the inequality is 906″. Cauchy's investigation is more fully elaborated by M. Puiseux in Annales de l'Observatoire de Paris, Vol. VII. In Vol. VIII, *ibid.*, Hoüel has recomputed the inequality.

The development of the reciprocal of the distance was later extended by Tisserand.[14] He shows that the development depending upon the inclination and eccentricity is divergent in some parts of the orbit of Pallas and proceeds to give the analytical development and to apply it to the case of Pallas.

In Bulletin Astronomique Vol. XII, 1895, M. P. Bruck has published the results of his work on "The secular variations of the elliptic elements of Pallas due to the action of Jupiter." He used the method developed by Gauss and extended by Hill and Callandreau. He utilizes elements by Farley for the epoch 1878.

In 1910 George Struve[15] published his results on "Die Darstellung der Pallasbahn durch die Gauss'sche Theorie für den Zeitraum 1803 bis 1910." The result of his work based on 63 normal places is a more accurate value for the mean motion of Pallas (769″.1385). The new value for the annual motion of Pallas compared with Jupiter becomes $18n' - 7n = 123''$. The deviation between observation and computation still amounts to $\pm 4'$ which is attributed to the second order perturbations. These residuals are somewhat reduced by empirical terms.

In A. N. No. 205, p. 225, M. Viljev has published his "Recherches sur le mouvement de Pallas." He attempts to reduce the residuals from Struve's work ($\pm 4'$) by taking into account second order terms in the general perturbations, employing the method by Hill. He reports his results as negative.

REFERENCES

[1] Bode's A. J. 1805, p. 102.
[2] Bode's A. J. 1805, pp. 106, 111, 228.
[3] Bode's A. J. 1805, pp. 181, 182.
[4] Bode's A. J. 1806, pp. 179–180.
[5] Gauss Werke. vol. vi, pp. 3–24.
[6] Gauss Werke. vol. vii, pp. 413–610.
[7] Bode's A. J. 1828, pp. 154, 157.
[8] Bode's A. J. 1829, p. 158.
[9] B. J. 1837, pp. 249–250.
[10] B. J. 1839, pp. 237–238. A. N. vol. 14, p. 329.
[11] B. J. 1851, pp. 547 and 549. A. N. vol. 55, p. 194.

[12] British Nautical Almanac 1860 (not available here).
[13] Tisserand Méchanique Céleste, vol. iv, p. 278.
[14] Annales de l'Observatoire de Paris. vol. xv.
[15] Dissertation Berlin. Ebernig, 1910 (not available here).

Additional references:

A. J. B., 1910, p. 188.
B. A. 28, p. 184.

TABLE 2.—Elements—(2) Pallas

Letter	Epoch	M.T.	L. (° ′ ″)	ϖ (° ′ ″)	Ω (° ′ ″)	i (° ′ ″)	φ (° ′ ″)	μ (′)	Authority	Equinox	Remarks
A....	1802 March 31.	Seeberg......	162 55 6.8	121 38 42	172 26 31	31 36 59	14 06 58	769.7263	Gauss	....	Gauss' Vth set Elements.
B.....	1802 Apr. 0....	....	162 51 14.5	122 03 02	172 28 57	34 50 40	14 15 31	760.98	Burkhardt	....	
C.....	1803.0.........	Seeberg......	221 28 54	121 24 13	172 28 08	34 38 20	14 13 06	769.416	Gauss	....	Gauss' VIth set of Elements.
D....	1803.0.........	Göttingen....	221 34 54	121 08 09	172 28 12	34 37 28	14 10 00	770.5010	Gauss	....	Based on oppositions 1803 to 1809 incl.
E.....	1803 156d.....	Göttingen....	254 56 38	121 07 54	172 28 21	34 37 44	14 12 46	770.7360	Gauss	....	Special perturbations
F.....	1810.0........	Göttingen....	47 23 05	121 08 05	172 31 44	34 36 09	14 02 50	769.1507	Gauss	....	Mean elements.
G.....	1812 June 10...	Göttingen....	239 04 46	121 00 48	172 32 44	34 34 55	13 59 02	768.5746	Gauss	....	
H.....	1826 June 24.5.	Paris.........	254 21 34	120 58 08	172 37 10	34 35 56	14 06 13	769.4911	Gauss-Encke	June 24	No statement regarding elements by Encke.
I.....	1838 June 21...	Berlin........	109 14 25	121 45 35	172 38 31	34 38 31	13 49 49	768.7126	Encke	Epoch	Last set of osculating elements by Encke.
J.....	1810 Jan. 0....	Göttingen....	49 05 14	121 14 08	172 33 37	34 37 43	14 12 51	770.7388	Galle	Epoch	
K.....	1850 Aug. 23...	Berlin........	338 52 59	121 21 48	172 44 00	34 37 33	13 52 32	768.4508	Galle	Epoch	Last set of Galle's osculating elements in A. N. 636.
L.....	1830 Jan. 0.0..	Berlin........	169 06 05	121 00 18	172 39 05	34 35 47	14 03 22	770.7516	Galle-Günther	1830	A. N. vol. 55, p. 194.
M....	1869 Sept. 25..	Berlin........	29 31 25	121 44 47	172 46 08	34 42 19	13 51 04	769.5682	Farley	Epoch	First set published by B. A. J. 1871.
N.....	1913 May 5....	Berlin........	193 37 07	121 57 36	172 56 48	34 42 02	13 46 38	769.2236	Farley	Epoch	Published in B. A. J. 1917.

(3) JUNO

Juno was discovered by Harding at Lilienthal near Bremen, September 1, 1804. Gauss computed several orbits successively correcting the elements by new observations. Elements VII[1] corrected with Bessel's observations, 1807. Ephemeris for 1808, April-December, approximately given. (Elements A.) A number of additional orbits were computed by Gauss' students at Göttingen (Wachter, Möbius, etc.).

Wachter[2]: Elements, (eccentricity omitted) from the last four oppositions after Gauss' Method, (Neue Comment. der Göttingen K. Societät, Bd. 1) including the opposition 1812. Eccentricity supplied from Bode's Astronomische Jahrbuch, 1816, p. 233. (Elements B.)

Möbius[3]: Oppositions used: 1810, 1811, 1812, 1813. Correcting mean longitude by $+4'55''$, the representation of the observations 1815, March, is $+8''$ in longitude, and $-51''$ in latitude. (Elements C.)

Nicolai[4] at Seeberg, near Gotha, compared Gauss' observations with the orbit of Möbius, (empirically correcting L), determined the oppositions and derived new elements. Oppositions used: 1811, 1812, 1813, 1815. "Juno is nearly in conjunction with Jupiter and the perturbations may be large." (Elements D.)

Taking up the determination of the large perturbations by Jupiter by the method of special perturbations, Nicolai[5] derived a new set of elements. Oppositions used: 1811, 1812, 1813, 1815, 1816, 1817, 1818. Representation of observations in 1819, $\Delta a + 2'.6$ $\Delta \delta$ $-0'.2$. (Elements E.)

Not satisfied with the representation of the observations by his last set of elements, Nicolai[6] extended the computation and determined new elements, ·which represented the observations of the "Atom" well in 1820. Oppositions used: 1805-1819. Special perturbations by Jupiter. Representation of the observations 1820, May, Δa $-7''$, $\Delta \delta$ $-2''$. (Elements F.) This computation of the special perturbations was continued for some years.

In 1823, Nicolai[7] derived his final set of elements, including the determination of the Jupiter mass, for which he found 1/1053.924, in agreement with the value Gauss had found from his theory of the motion of Pallas. The representation of the observations cannot be improved by taking Saturn or Mars into consideration, but Nicolai considers the possibility of the active mass of Jupiter changing with the body acted upon. From these elements osculating elements for 1826 were computed, taking account of the special perturbations by

Jupiter. Berliner Jahrbuch uses the elements by Nicolai until 1830. Fifteen oppositions used: 1804-1823. Special perturbations by Jupiter, (Saturn, Mars, negligible). Residuals in longitude —23″ to +27″, still show a run with the period of Jupiter. (Elements G.)

In 1832 new elements by Encke[8] were introduced, and are carried forward with special perturbations to 1865 by Bremiker and Powalky, for the ephemerides published 1832-1865. Perturbations by Jupiter with mass, 1/1053.924. (Elements H.)

Damoiseau[9] has published general perturbations in the Connaissance des Temps.

Hind[10] took over the work started by Nicolai, Encke, and Bremiker, to compute osculating elements for each opposition by special perturbations. The ephemerides are published in the Nautical Almanac. and the Berliner Jahrbuch. As a basis for this work he derived new elements. (Elements I.)

An attempt to apply Hansen's method of determination of the general perturbations was made by Berkiewicz,[11] starting with Hind's elements. The perturbations of the first order with regard to Jupiter, Mars, and Saturn were determined, also the constants of integration leading to a mean motion, 814″.090. No comparison with the observations is attempted.

Being aware that the corrections to the ephemerides computed according to Hind had increased to 3′ in 1887, Downing[12] undertook to correct Hind's elements. The errors of the tabular heliocentric places published in Greenwich Observations, 1864-1887, are discussed. Equations for the longitude and latitude corrections were set up expressed in terms of corrections to the elements and combined to eliminate the corrections to the radius vector. The mean motion is included in the solution and receives by far the greatest weight. The representation of the oppositions show a pronounced run in Δa. These elements were used for the computation of the annual ephemerides to 1913 in the Nautical Almanac and to 1917 in the Berliner Jahrbuch. (Elements J). Oppositions used: 1864-1887. Special perturbations by Venus, the Earth, Mars, Jupiter, Saturn. Representation varies from —3″ to +4″ in $\Delta a \cos \delta$, and —1″ to +2″ in $\Delta \delta$. Large residual (—11″) for 1874. Representation for 1890 is then in $a+3″$, in $\delta \pm 0″$ against —65″ and —6″, according to Hind's computations.

Since 1917 Ephemeriden der Kleinen Planeten gives mean elements by Boda[13] derived by the method of Brendel. (Elements K.) Mean elements. Perturbations by Jupiter according to Brendel, A. N., Vol. 195, p. 417. Expected representation ±0°.5 to year 2000. Oppositions not stated.

The absolute perturbations according to the method of Hansen (?), have been computed by Viljev.[14]

REFERENCES

[1] Bode's B. J. 1811, p. 136. Gauss Werke Bd. 6.

[2] Bode's B. J. 1815, p. 248.

[3] Bode's B. J. 1817, p. 213.

[4] Bode's B. J. 1818, p. 264.

[5] Bode's B. J. 1820, p. 200.

[6] Bode's B. J. 1822, p. 218.

[7] Bode's B. J. 1826, p. 224.

[8] A. N. Bd. 27, p. 177.

[9] Connaissance des Temps, 1846. Additions. Not available here.

[10] Nautical Almanac, 1859. Appendix. Not available here.

[11] A. N. vol. 72, p. 1, p. 145, p. 289.

[12] N. M. vol. 50, p. 487.

[13] A. N. vol. 200, p. 1.

[14] Bulletin-Soc. Astr. Russ. vol. 22. Not available here.

TABLE 3.—*Elements—(3) Juno*

Letter	Epoch	M. T.	L a M.	π	Ω	i
			° ′ ″	° ′ ″	° ′ ″	° ′ ″
A.....	1805.............	Göttingen.....L	42 37 3.7	53 19 0.2	171 4 28.2	13 4 26.2
B.....	1811.............	Göttingen.....L	177 48 21.0	53 15 10.1	171 9 16.7	13 4 17.2
C.....	1810............,	Göttingen.....L	95 29 53.2	53 6 43.0	171 6 45.0	13 4 12.9
D....	1815 Dec. 31.0....	Göttingen.....L	230 11 34.2	53 14 53.8	171 9 58.9	13 4 0.1
E....	1819.............	Mannheim....L	117 45 2.84	53 32 56.09	171 6 50.23	13 3 37.29
F....	1820 May 11......	Mannheim.....L	230 9 22.03	53 31 6.52	171 8 11.08	13 3 47.20
G....	1810.............	Göttingen.....L	95 25 9.82	52 58 35.89	171 6 28.52	13 4 18.99
H....	1826 Nov. 1......	Berlin........M	351 43 27.3	53 11 13.4	170 56 57.4	13 3 28.4
I.....	1861 Nov. 21.0...	Greenwich....L	58 34 1.0	54 9 3.3	170 59 49.7	13 2 58.8
J.....	1861 Nov. 21.0...	Greenwich.....	58 34 1.83	54 9 3.82	170 5 45.69	13 2 58.45
K....	1900 Jan. 0.......	L	324 12	55 36	170 42	13 2

Letter	Epoch	M. T.	φ	μ	Equinox	Authority
			° ′ ″	″		
A.....	1805...........	Göttingen.......	14 48 11.5	813.8468	....	GaussVII
B.....	1811...........	Göttingen.......	14 44 1	813.25748	1811	Wachter
C.....	1810...........	Göttingen.......	14 43 9.5	812.7140	1810	Möbius
D....	1815 Dec. 31.0..	Göttingen.......	14 43 28.84	812.9304	1816.0	Nicolai
E....	1819...........	Mannheim......	14 53 17.44	813.86981	1819	Nicolai
F.....	1820 May 11....	Mannheim......	14 55 1.78	814.40238	1820 May 11	Nicolai
G....	1810...........	Göttingen.......	14 44 39.19	813.4837354	1810	Nicolai
H....	1826 Nov. 1....	Berlin........	14 53 22.6	813.88514	1810	Encke
I....	1861 Nov. 21.0..	Greenwich.......	14 47 14.1	813.34555	1861 Nov. 21.0	Hind
J.....	1861 Nov. 21.0..	Greenwich.......	14 47 13.81	813.35271	1861 Nov. 21.0	Downing
K....	1900 Jan. 0.....	....	14 50	813.434	1900	Boda

(4) VESTA

Vesta was discovered by Olbers[1] at Bremen on March 29, 1807.

Preliminary elements were computed by Gauss[2] and also by Burkhardt.[3] The third set of Elements A by Gauss is based on observations from March 29 to July 11. They were used to compute the ephemeris for 1808-1809.

The preliminary work of Gauss was continued by Gerling,[4] who supplied the ephemeris for a number of years. His last set of Elements B, are based on the first six oppositions.

Burkhardt's preliminary work was continued by Daussy[5] (reference not available here). In his work he took into consideration the perturbations by Jupiter, Saturn, and Mars, and was able to represent the first seven oppositions satisfactorily. On account of the small eccentricity and inclination, the methods of La Place and Le Verrier were sufficient.

On account of increased error in the ephemeris for 1818 based on Gerling's first elements, Encke[6] computes a set of Elements C, based on oppositions 1812, 1815, 1816, and 1818.

In Astronomisches Jahrbuch 1829, pp. 156-158, Encke gives the results of his work on Vesta based on fourteen oppositions by also using Nicolai's value of Jupiter's mass (1/1054), for the perturbations due to Jupiter. He also makes use of Daussy's tables for the perturbations of Saturn and Mars. His new set of Elements D represents the oppositions from 1807 to 1825 within $\pm 6''$. Elements D are then brought up to the epoch of 1827. Astronomisches Jahrbuch from 1830 to 1866, contains the elements and ephemerides computed by Encke. (See Elements E.) The method of computation[7] was that of the variation of the elements by special perturbations of Jupiter. In this work Encke was assisted by Bruhns and Schiaparelli. In A. N. 332, Encke comments on the poor results obtained for planets (1), (2) and (3), by using Laplace's value for the mass of Jupiter and also publishes a new value of Jupiter's mass (1/1050), obtained from the results of Vesta.

The Berliner Jahrbuch for 1868 publishes mean Elements F by Brünnow.[8] They are also published in Watson's Theoretical Astronomy. Brünnow completes the work of Wolfers and Galle[9] who developed expressions for the perturbations in longitude and radius vector after Hansen's method. Brünnow's elements represent the oppositions from 1810 to 1851 within $-6''$ to $+10''$. For Jupiter's mass he used 1/1050.

From 1871 to 1910 the Berliner Jahrbuch publishes elements (see Elements G), by Farley.[10] His work is based on twelve oppositions, 1840 to 1855. This work forms the basis for later investigations of the general perturbations.

Probably the most extensive work on a minor planet are the tables of Vesta by Leveau published in Annales de l'Observatoire de Paris Mémoires, XV, XVII, XX, XXII, XXV. The method applied by Leveau is that of Hansen "Auseinandersetzung einer Zweckmässigen Methode zur Berechnung der Absoluten Störungen der Kleinen Planeten, I, II, III." The explanation for the choice of this method is that the application to Vesta is a preparatory study to the motion of Pallas, as Gauss' theory of Ceres was a preliminary study to his theory of Pallas. Memoir XV contains the perturbations of the first order of the masses of Venus, Earth, Mars, Jupiter, Saturn, Uranus, and Neptune. The memoir concludes with the determination of the constants of integration and the expressions for $n\delta z$, v, u seci and a representation of an observation 1858, April, 23.5 as follows: $\Delta a = -0^s.1$ and $\Delta \delta = -0''.4$. Memoir XVII contains the terms depending upon the square of the mass of Jupiter in $n\delta z$ and $2v$. The effect on u seci becomes noticeable after 100 years. Memoir XX contains the terms depending upon the product of the masses and concludes with a set of mean elements and corresponding expressions for $n\delta z$, v, and u seci. The mean elements given in memoir XX are slightly changed in memoir XXII (see Elements H), on account of some perturbations of the second order that Farley had included and which form the basis of Leveau's work. Memoir XXII contains the comparisons with 215 normal places founded on 5000 observations extending from 1807 to 1889. The computation has been changed to conform to the solar tables by Newcomb. The correction to the mean motion is zero. The new Elements I represent the observations in right ascension between $-2''$ and $+4''$ and in declination between $-1''$ and $+4''$. A new value of Jupiter 1/1046, and Mars 1/3648000. The perturbations are collected in tables introducing the mean anomaly and corrective terms for the eccentric anomaly. Memoir XXV contains some supplementary terms depending upon the product of the masses. One of the larger terms has a period of 3000 years. The effect of the critical terms, which are mentioned in the beginning of his work, shows itself by comparing these terms before and after integration. The coefficients of corresponding terms are ten or twenty times larger, whereas the other terms mostly decrease.

Leveau has made a comparison between observations and calculation of later positions[11] showing the residuals as obtained from the British Nautical Almanac, and his own computations. The following comparisons are illustrations of the results:

		1890	1892	1894	1896	1898
Nautical Almanac	$\Delta\alpha$ +1^s.18	+1^s.00	+1^s.73	+1^s.71	+2^s.91
	$\Delta\delta$ +0″.9	—5″.8	+8″.9	—2″.0	+16″.3
Leveau	$\Delta\alpha$ +0^s.03	+0^s.01	+0^s.25	+0^s.06	+0^s.19
	$\Delta\delta$ +0″.4	+0″.5	+2″.0	+1″.5	+1″.0

Further results of Leveau's theory are published in Comptes Rendus. T. 145, p. 903-906, "Détermination des Éléments Solaires et des Masses de Mars et de Jupiter par les Observations Méridiennes de Vesta." Extending the comparison with the meridian observations from 1807 to 1904 and taking into consideration the masses of Jupiter and Mars and also the solar elements, Leveau determines a new set of smaller corrections to the elements of Vesta and for Jupiter's mass, 1/1046, and mass of Mars 1/3601280. The tables of residuals shows the poor quality of the meridian observations before 1826. A period of 36 years, (three revolutions of Jupiter, or ten of Vesta), points to the effect of the critical terms in the residuals; the amplitude is about 1″. The effect of the earlier observations on the value for the mean motion is also illustrated by the last residuals.

In Annales de l'Observatoire Astronomique de Toulouse, T. I., B. 1 to B. 90, M. J. Perrotin published his extensive investigation on the "Théorie de Vesta," applying the method of Le Verrier (Annales de l'Obs. de Paris, T. X.). The method consists first of deriving mean elements from previous osculating elements by computing provisional periodic perturbations and applying these to the osculating elements for a first approximation. In order to avoid considerable labor, Perrotin starts with a certain fixed major axis and develops corrective terms for the variation in the assumed value. The final mean motion is determined from two extreme groups of observations in 1807 and 1876 when the planet was near the same place in its orbit. The perturbative function is developed by Le Verrier to the seventh degree in the inclination and eccentricity; thus Perrotin includes terms of $10n'$— $3n$. Venus, Earth, Mars, Jupiter, and Saturn are taken into account. Derived from the secular terms e is always smaller than 0.15, the mean motion of the perihelion is $+38''$, that of the node $—38''$, and the inclination remains less than 9°. The terms of the second order are then considered. Those due to the square of Jupiter's mass of the second degree are small. Those depending on the product of the masses are more important, especially those depending upon $5n''—2n'$,

$-2n''+4n'-n$, $2n''+9n'-3n$. For getting these terms the development of the perturbative function is used to the seventh degree and for determining the terms of the eighth degree the method of Cauchy, as extended by Puiseux, is used. No comparison with observations is attempted.

REFERENCES

[1] Bode's B. J. 1810, p. 194,

[2] Bode's B. J. 1810, pp. 198, 213; 1812, p. 253; Gauss Werke. vol. vi.

[3] Bode's B. J. 1810, p. 199; Annales de l'Observatoire Astronomique de Toulouse. vol. i, p. B4.

[4] B. J. 1814, p. 253; B. J. 1817, p. 255; B. J. 1819, p. 224.

[5] Connaissance des Temps, 1818, 1819, 1820.

[6] Bode's B. J. 1821, p. 220.

[7] B. J. 1838, p. 287 to 293.

[8] Astr. Notices. vol. i.

[9] Mem. Berlin Academy, 1841.

[10] British N. A. 1860.

[11] Bull. Astr. vol. 19, p. 434; Comptes Rendus, T. 135, p. 525.

TABLE 4.—Elements—(4) Vesta

Letter	Epoch	L (° ′ ″)	M (° ′ ″)	ω (° ′ ″)	Ω (° ′ ″)	i (° ′ ″)	φ (° ′ ″)	μ (″)	Authority	Remarks
A.....	1807 March 31, Bremen.	192 23 30	....	146 32 04	103 18 28	7 08 11	4 54 18	981.7087	Gauss	3rd set of preliminay orbits.
B.....	1816.0, Göttingen......	341 07 43	....	146 29 06	103 13 29	7 08 10	5 07 56	977.8933	Gerling	Based on first six oppositions.
C.....	1818.0, Seeberg	179 38 30	....	146 20 06	103 11 38	7 08 09	5 06 24	977.7020	Encke	
D.....	1810.0, Paris...........	105 53 16	216 04 49	146 40 06	103 08 20	7 08 12	5 09 39	978.2967	Encke	
E.....	1831 July 23, Berlin....		105 35 26	145 51 09	103 20 28	7 07 57	5 04 51	977.7554	Encke	B. J. 1831, p. 249.
F.....	1810 Jan. 0, Berlin.....		216 42 26	146 08 07	103 11 22	7 08 05	5 05 36	977.6339	Brünnow	
G.....	1869 May 13, Berlin.....		345 18 50	146 38 57	103 26 23	7 07 53	5 07 42	977.7104	Farley	B. J. 1871.
H.....	1857 Jan. 1.0, Paris....		198 20 17	147 10 59	103 23 19	7 08 07	5 06 05	977.6324	Leveau	Mem. XXII, p. A1.
I.....	1857 Jan. 1.0, Paris....		198 20 33	147 10 40	103 23 20	7 08 06	5 06 04	977.6325	Leveau	Mem. XXII, p. A57.

(10) HYGIEA

De Gasparis at Naples announced to the General-Secretary of Corrispondenza Scientifica that he had discovered this planet April 12, 1849. The magnitude was 9–10.[1]

A large number of preliminary orbits were computed by Encke,[2] Luther,[3] Brorsen,[4] Quirling,[5] d'Arrest,[6] Hensel,[7] Santini,[8] from observations in the first opposition with mean motions ranging from 601″ to 652″. The first more certain, Elements A, were those by d'Arrest[9] from observations in 1849 and 1850. These elements were published in B. J. 1853–55, brought forward with special perturbations by Jupiter.

In B. J. for 1856 two sets of elements were published: one by Chevallier at Durham, B, from observations in 1851 and 1852; and one, C, by Zech at Tübingen. The latter (corrected for an error[10] which did not affect the ephemeris) were based upon four oppositions and represented the opposition 1854 by $+0^s.5$ in α and $-1″$ in δ. This was the beginning of Zech's important work which made this planet suitable for testing theories in the case of near commensurability.

In B. J. 1858, new Elements D, by Zech,[11] are published, based upon five oppositions taking the perturbations by Jupiter, Saturn, and Mars into account. These elements are very closely the same as those v. Zeipel selected from Zech's manuscript. In this the general perturbations by Jupiter, Saturn, and Mars were computed and partly tabulated. The basic elements for this computation were founded on eight oppositions. After Zech's death in 1864 his computations of the general perturbations of the planets (5) and (10) were discontinued by the Recheninstitut, whereas his elements and special perturbations for (10) Hygiea were used in B. J. until 1875 by Powalky and Becker. In 1873 the correction to the ephemeris had increased to -4^s in α and $-20″$ in δ.

In B. J. 1876 E. Becker published a preliminary set of elements which finally was corrected to the Elements E, given by Bauschinger[12] based upon the oppositions 1868, 1869, 1871, 1873 and 1874. Special perturbations by Jupiter and Saturn were included and brought forward to the osculation 1898, December 20.

The elements given in Kleine Planeten for 1920 are Becker's brought forward by Strehlow with Jupiter's perturbations. The mean motion is corrected empirically by $+0″.07$ (since 1898, August 22), representing the fourteen oppositions since 1900 within $\pm0^m.2$. Osculation 1920 January 0. Elements F.

In the mean time v. Zeipel[13] had computed the tables for the general perturbations by Jupiter according to the method by Bohlin in the case of near commensurability, Hecuba group. As a test he applied his tables to the orbit of (10) Hygiea and started with the Elements G (a) from Zech's manuscript. These were first transformed to mean elements and then compared with nine oppositions from 1849–1884 with the aid of his tables. A least squares solution gave the corrected Elements H.

The work of computing tables after Bohlin's method for the Hecuba group had also been undertaken by A. O. Leuschner[14] and an application to (10) Hygiea was made by Miss E. Glancy and Miss S. H. Levy[14]. In order to compare the results with those of v. Zeipel, Miss Glancy[15] used the Berkeley tables and after a comparison with nine oppositions between 1849–1884 obtained the Elements I, starting with the mean Elements G(b). Further comparisons were made by representing observations in 1910, 1914, 1917. The residuals before solution from Zech's elements with the Berkeley tables were —11' to +10' in the plane; from Zech's elements and v. Zeipel's tables —24' to +5'. After Miss Glancy's solution the residuals are —8' to +7', and after v. Zeipel's solution —7' to +8'; but in 1917 they are +9' and +19' respectively, apparently in favor of the Elements I and the Berkeley tables.

The residuals from Zech's elements and the Berkeley tables (—11' to +10') show a decided periodicity of 30 years, thus pointing to the influence of Saturn. Later observations in 1917 and 1921 are represented much better (0' and +10') by the Berkeley tables and Zech's elements than by any of the solutions.[16] The best future representation may be expected from Zech's elements G(b) and the Berkeley tables. The residuals, probably chiefly due to Saturn, keep within fixed limits ±10'. The most obvious next step would be to correct the residuals for some of the earlier oppositions by means of the perturbations of Saturn and Mars, available in manuscript in the Rechen-institut. Until that shall have been done, no corrections should be applied to Zech's elements, which appear to be the best available.

REFERENCES

[1] A. N. vol. 28, p. 391.
[2] A. N. vol. 29, p. 15.
[3] A. N. vol. 29, p. 49.
[4] A. N. vol. 29, p. 81.
[5] A. N. vol. 29, p. 81.
[6] A. N. vol. 29, p. 81, p. 126; vol. 30, p. 320.
[7] A. N. vol. 30, p. 81, p. 82.
[8] A. N. vol. 30, p. 87.
[9] A. N. vol. 31, p. 276.
[10] A. N. vol. 38, p. 327.
[11] A. N. vol. 39, p. 347.
[12] Veröffentlichungen des astronomischen Recheninstituts zu Berlin. Nr. 16, p. 48.

[13] H. v. Zeipel, Angenäherte Jupiterstörungen für die Hecuba-Gruppe. Mémoires de l'Académie des Sciences de St. Petersbourg, vol. 12, Nr. 11, 1902.
[14] National Academy of Sciences. Memoirs, vol. 14, third memoir.
[15] A. J. vol. 32, p. 27.
[16] A. O. Leuschner, Comparison of theory with observation for the minor planets (10) Hygiea and (175) Andromache. Proc. N. A. S., Washington, vol. 8, No. 7, p. 170.

TABLE 5.—Elements—(10) Hygiea

Letter	Date	M.T.	M	π	Ω	i
			° ′ ″	° ′ ″	° ′ ″	° ′ ″
A.....	1849 Apr. 15.0..	Berlin......	330 52 8.56	227 49 54.23	287 37 8.64	3 47 15.51
B.....	1851 Sept. 28.5..	Berlin......	128 44 20.7	228 2 32.1	287 39 8.3	3 47 8.3
C.....	1851 Sept. 17.0..	Berlin......	126 59 37.2	227 48 9.2	287 38 37.6	3 47 9.22
D.....	1851 Sept. 17.0..	Berlin......	126 59 48.76	227 47 58.77	287 38 34.21	3 47 9.29
				ω		
E.....	1874 Dec. 26.0..	Berlin......	174.55.30.0	312 40 30.5	285 18 57.5	3 47 43.2
F.....	1925 Jan. 0.0..	Greenwich..	181 38 49.2	305 25 22.8	285 52 55.2	3 48 46.8
				π		
G (a)..	1851 Sept. 17.0..	Berlin......	126 59 48.6	227 46 36.6	287 37 11.4	3 47 8.4
G (b)..	1851 Sept. 17.0..	Berlin......	121 51 58	230 47 49.6	287 37 11.28	3 47 8.5
H.....	1851 Sept. 17.0..	Berlin......	121 35 27.6	231 2 9	287 8 33.6	3 47 45.6
I......	1851 Sept. 17.0..	Berlin......	121 31 53.8	231 4 56.6	287 27 28.1	3 47 30.1

Letter	Date	M.T.	φ	μ	Equinox	Authority
			° ′ ″	″		
A.....	1849 Apr. 15.0.....	Berlin....	5 47 55.37	634.6406	1849.0	d'Arrest VI
B.....	1851 Sept. 28.5.....	Berlin....	5 46 34.7	634.83504	....	Chevallier
C.....	1851 Sept. 17.0.....	Berlin....	5 46 16.8	634.84564	1851 Sept. 17	Zech
D.....	1851 Sept. 17.0.....	Berlin....	5 46 16.57	634.84912	1851 Sept. 17	Zech
E.....	1874 Dec. 26.0.....	Berlin....	6 18 23.7	636.58673	1870.0	E. Becker
F.....	1925 Jan. 0.0...	Greenwich	6 40 48.0	638.517	1925.0	Strehlow
G (a)..	1851 Sept. 17.0.....	Berlin....	5 46 16.8	634.850	1850.0	Zech
G (b)..	1851 Sept. 17.0.....	Berlin....	6 23 8.9	636.8566	1850.0	Zech
H.....	1851 Sept. 17.0.....	Berlin....	6 22 1.2	636 849	1850.0	v. Zeipel
I......	1851 Sept. 17.0.....	Berlin....	6 21 31.0	636.86105	1850.0	Glancy

(28) BELLONA

Discovered by R. Luther[1] at Bilk near Düsseldorf, March 1, 1854.
Preliminary elements were published by Bruhns,[2][3][4] Chevallier[5] and Ruemker,[5] Oudemans.[6]

From 141 observations formed into five normal places Bruhns[7][8] derived, originally using four longitudes and two latitudes corresponding to the first, second, fourth and fifth normals, the first reliable Elements A. As Elements A differ considerably from his previous elements and those of Oudemans. He made a comparison of ephemerides which satisfied him regarding the correctness of Elements A. This case is somewhat indeterminate and small errors of observation would produce considerable changes in resulting elements. From Elements A Bruhns[7] has published an ephemeris for 1855.

Further ephemerides are published by Bruhns, among them for 1856,[9] 1866,[10] (a star correction[11] November 29 by Engelmann), for 1867,[12] correction Δa —50[s], $\Delta \delta$ —3'.4 by Tietjen,[13] and for 1871–72.[14]

Other elements by Bruhns are published in the B. J. from 1857 to 1860.

The B. J. from 1861 to 1891 contains new elements and ephemerides by Bruhns originally based upon the observations of the first four oppositions with perturbations by Jupiter, Saturn, and Mars. Elements C.[19] Whether these elements are merely brought forward by perturbations or contain corrections is difficult to determine.

From 1892 the B. J. uses von der Groeben's elements instead of those by Bruhns.

By a process of successive correction of osculating elements and special perturbations by Jupiter, Saturn and Mars (perturbations by the Earth and Venus were found negligible), based on 16 normal places extending over a period of thirty-two years from 1854 to 1886, von der Groeben,[15] starting with a set of elements osculating for 1870, September 18, derived Elements Ba, Bb, Bc, Bd osculating for different epochs from 1861 to 1882. These elements were brought forward with special perturbations of Jupiter, Saturn and Mars to the epoch 1886, February 26, Elements Be, and to the epoch 1889, October 28, Elements Bf. Elements Bf are adopted by the B. J. for 1892[16] and 1893.[17] Three observations by Ball at Lüttich, October 31 to November 15, 1889, are well represented by the ephemeris. Mean Δa +[s].19, mean $\Delta \delta$ —".5.

From seven observations at Washington and Tacubaya with star places newly determined by Bruns and four observations by himself at Düsseldorf, Luther[18] obtained a mean correction to the ephemeris from von der Groeben's Elements Bf brought forward by special perturbations of $\Delta\alpha$ —$0^s.42$, $\Delta\delta$ +$3''.4$ in 1894.

The Elements Bg[20] to Bu given in the B. J. and in Kleine Planeten from 1894 to 1919 are von der Groeben's elements probably brought up to date in the same manner as before with the special perturbations by Jupiter, Saturn and Mars and without any other correction.

The following is a partial list of published corrections to the B. J. ephemerides from von der Groeben's elements:

1902, August 7	$+\quad 8^s$	$+\ 0'$	Luther[21]
1903, October 26	$+6^m38$	$+25'.2$	Luther[22]
1905, March 1	$+\quad 2.77$	$-\ 8'.3$	Iwanowski[23]
1906, June 20	$+\quad 8.99$	$+11.3$	Luther[24]
1907, September 7	$+\quad 21.68$	$+1'\ 23'.1$	Luther[25]
1908, November 28	$-\quad 17.77$	$+0'\quad 9'.0$	Luther[26]
1916, August 8	$+0^m.3$	$+1'$	Luther[27]
1917, December 4	$+1^m4$	$+7'$	Luther[28]
1919, April 1	-1^m8	$+6'$	Luther[29]

In a dissertation on the Jupiter perturbations of the group of small planets whose mean daily motions are in the neighborhood of $750''$, D. T. Wilson[30] gives an application of the Hansen-Bohlin method to the Jupiter perturbations of this group. "The integration divisors for certain values of the integers n, r and s become as small as 0.2. These terms increase rapidly as the series advance. They were computed to the third power of the eccentricities and to the fourth power of w. It was found that all the terms of the third and fourth powers of ψ and some of those of the third power of the eccentricities are negligible when the eccentricity of the disturbed planet does not exceed 0.34 and when the mean daily motion lies within the limits $720''$ to $780''$. Therefore only those terms of the third power of the eccentricities which are appreciable within the above limits have been retained. All the secular terms have been computed to the fourth power of eccentricities."

By means of these tables the Jupiter perturbations of Bellona were computed and compared with the results previously obtained by Hansen's method by Bohlin.[31]

The mass of Jupiter is taken as 1: 1048 in the tables by Wilson.

Of the three applications of these tables by D. T. Wilson that of Bellona is by far the most interesting. This depends on the greater proximity to the commensurability ($748''$) and the cross position of the line of apsides to that of Jupiter. The inequality 2g–5g′ in $\eta\delta z$ is the largest of all and amounts to $40'$. But the difference between the coefficients computed by Hansen's and Bohlin's methods is large ($2'$) and the comparison with the observations ought to decide between the application of these methods to numerous planets in this group.

REFERENCES

[1] A. N.   vol. 38, pp. 111, 143.
[2] A. N.   vol. 38, p. 155.
[3] A. N.   vol. 38, p. 218.
[4] A. N.   vol. 38, p. 351.
[5] A. N.   vol. 38, p. 158.
[6] A. N.   vol. 38, p. 158.
[7] A. N.   vol. 40, p. 203.
[8] B. J.   1858, p. 407.
[9] A. N.   vol. 44, p. 235.
[10] A. N.   vol. 68, p. 125.
[11] A. N.   vol. 69, p. 104.
[12] B. J.   1869.
[13] A. N.   vol. 65, p. 176.
[14] A. N.   vol. 79, p. 5.
[15] A. N.   vol. 123, p. 369.
[16] B. J.   1892, p. 390.
[17] B. J.   1893, p. 390.

[18] A. N.   vol. 139, p. 302.
[19] B. J.   1861, p. 506.
[20] B. J.   1894, p. 394.
[21] A. N.   vol. 159, p. 295.
[22] A. N.   vol. 163, p. 383.
[23] A. N.   vol. 167, p. 319.
[24] A. N.   vol. 171, p. 349.
[26] A. N.   vol. 175, p. 401.
[26] A. N.   vol. 179, p. 243.
[27] A. N.   vol. 203, p. 164.
[28] A. N.   vol. 206, p. 16.
[29] A. N.   vol. 208, p. 248.
[30] Astronomiska Iakttagelser och Undersökningar a Stockholms Observatorium. vol. 10, No. 1.
[31] Manuscript in the office of the Recheninstitut.

TABLE 6.— Elements—(28) Bellona

Letter	Epoch	M. T.	M.	π	Ω	i	φ	μ	Equinox	Authority
A	1854 Mar. 0.0	Berlin	36 43 21	122 18 29	144 43 7	9 22 33	8 53 54	767.5226	1854.0	Bruhns
Ba	1861 Sept. 25.0	Berlin	264 55 34	338 10 21	144 39 21	9 21 26	8 38 11	765.9065	1860.0	v. d. Groeben
Bb	1866 Oct. 9.0	Berlin	296 28 59	338 20 32	144 38 29	9 21 38	8 47 4	766.8382	1870.0	v. d. Groeben
Bc	1870 Sept. 18.0	Berlin	243 24 36	338 6 45	144 37 17	9 21 47	8 50 59	767.3374	1870.0	v. d. Groeben
Bd	1882 Apr. 28.0	Berlin	64 43 25	339 18 14	144 37 35	9 21 31	8 34 10	765.8751	1880.0	v. d. Groeben
Be	1886 Feb. 26.0	Berlin	2 10 25	339 40 53	144 44 36	9 21 35	8 35 28	766.1202	1890.0	v. d. Groeben
Bf	1889 Oct. 28.0	Berlin	287 46 46	339 19 23	144 39 28	9 21 27	8 40 59	766.4180	1890.0	v. d. Groeben
Bg	1881 Feb. 20.0	Berlin	30 3 12	339 11 38	144 39 11	9 21 28	8 41 24	766.1111	1890.0	v. d. Groeben
Bh	1893 Nov. 16.0	Berlin	243 19 49	338 53 47	144 38 42	9 21 31	8 42 58	766.8942	1890.0	v. d. Groeben
Bi	1896 Apr. 14.0	Berlin	71 10 8	338 24 33	144 43 21	9 21 42	8 41 6	766.6272	1900.0	v. d. Groeben
Bj	1897 June 28.0	Berlin	164 45 50	338 30 4	144 43 20	9 21 42	8 40 6	766.2606	1900.0	v. d. Groeben
Bk	1898 Sept. 11.0	Berlin	258 21 44	338 30 59	144 43 16	9 21 37	8 38 55	765.9782	1900.0	v. d. Groeben
Bl	1900 Feb. 13.0	Berlin	8 58 37	338 31 49	144 42 22	9 21 38	8 38 00	766.2998	1900.0	v. d. Groeben
Bm	1903 Nov. 5.0	Berlin	295 51 7	340 41 44	144 34 37	9 23 12	8 38 1	766.7700	1900.0	v. d. Groeben
Bn	1905 Apr. 8.0	Berlin	46 31 26	340 48 4	144 42 37	9 23 4	8 39 10	766.4981	1910.0	v. d. Groeben
Bo	1906 June 22.0	Berlin	140 8 28	340 50 1	144 41 37	9 23 4	8 40 14	766.3312	1910.0	v. d. Groeben
Bp	1907 Sept. 5.0	Berlin	233 47 46	340 50 12	144 41 21	9 23 12	8 41 35	766.6662	1910.0	v. d. Groeben
Bq	1908 Dec. 28.0	Berlin	336 3 47	340 49 47	144 41 19	9 23 14	8 42 32	766.9046	1910.0	v. d. Groeben
Br	1910 Apr. 22.0	Berlin	78 25 24	340 38 24	144 40 13	9 23 1	8 45 7	766.6520	1910.0	v. d. Groeben
Bs	1911 July 6.0	Berlin	172 23 32	340 25 38	144 39 15	9 23 3	8 45 42	767.2816	1910.0	v. d. Groeben
Bt	1912 Oct. 28.0	Berlin	274 51 16	340 18 9	144 39 2	9 23 58	8 45 5	766.913	1910.0	v. d. Groeben
Bu	1925 Jan. 0.5	Greenwich	142.319	340.308	144.854	9.398	8.751	766.913	1925.0	v. d. Groeben
C	1857 Dec. 15.0	Berlin	331 41 53	122 24 28	144 38 58	9 21 24	8 39 0	766.1418	1854.0	Bruhns.

(93) MINERVA

Discovered by J. Watson [1] 1867, August 24, at Ann Arbor, and observed on three successive days. Estimated to be of 11th magnitude.

P. Lehmann [2] in Berlin computed his first orbit from the observations to October 2, covering an heliocentric arc from 60° to 75° after perihelion. An ephemeris for 1867, November and December, deviated $+0^s.33$ in a, $-2''.0$ in δ from the observations. These elements were repeated in B. J. 1870. Elements A.

Lehmann changed his Elements A to the Elements B given in B. J. 1871 without statement of the observations used, the perturbations applied or the representation of the observations. The most prominent changes are the improvement of the longitude of perihelion by 1°, and also the increase of the eccentricity, besides diminution of the mean motion by 1″ a day, to a value as much below the mean mean motion as the osculating value was above. This error in the mean motion made a new computation necessary which Lehmann published in B. J. 1872 without explanation. Elements C. A further improvement was obtained by the next set of Elements D by Lehmann in B. J. 1873. These elements were (perhaps) brought up with perturbations to the new osculation in 1872 as published in B. J. 1874 and 1875. Elements E. To the elements M, μ and ϕ corrections were applied. The resulting Elements F are given in B. J. for 1876-1880. Then Lehmann undertóok to make a final determination of the orbit from the first 7 oppositions with special perturbations by Jupiter and Saturn, (as stated in Bauschinger, Tabellen, etc.) These Elements G were probably published in B. J. with change of osculation until the issue 1913.

Berberich corrected the Elements G empirically by the observations in the oppositions 1899, 1902, 1907, 1908, 1911, and derived the Elements H, which were published in B. J. 1914. B. J. 1915 gives the elements by Leuschner.[6]

The general perturbations by Jupiter were developed by W. S. Eichelberger.[3] The basic elements were obtained "from a special discussion, by the author, of the observations from 1867 to 1879, inclusive." Elements I. The method is that of Hansen for absolute perturbations of the first order retaining the eccentric anomaly in the argument. The constants of integration were determined. With these perturbations and the preliminary Elements I the observations from 1867 to 1884 were compared. A least squares solution was made and the final Elements J obtained. The comparison with the observations is not given.

Observations by T. J. J. See in 1899 Sept. compared with a manuscript ephemeris by Eichelberger gave the corrections $+0^s.8$ in a, $+50''$ in δ. "The comparison would indicate that Eichelberger's theory is very good." Same year and date Coddington observed Minerva and compared "with an ephemeris furnished by· Professor Newcomb." Same corrections as See found to Eichelberger's ephemeris.

The planet discovery 1902 HQ (February 25), by Wolf[4] was suspected to be (93) Minerva and the identity was confirmed by two observations from Bordeaux.[5]

The work of Eichelberger was undertaken originally under the auspices of the Watson Trustees, but the Trustees suspected that the small residuals which resulted from a comparison of theory and observation were due to some error. Investigation of the sources of the suspected error undertaken by Leuschner at the request of the Trustees confirmed Eichelberger's work, the representation of the observations being found entirely satisfactory, in view of the fact that only first order perturbations of Jupiter were considered.

Leuschner[6] revised the elements by including in the least squares solution further oppositions to 1902, so that his elements are based on oppositions extending from 1867 to 1902. The elements differ only slightly from those by Eichelberger. In the Perturbations and Tables of Twelve Watson Asteroids,[6] Eichelberger's perturbations are retained without change. Observations of recent years are well represented by the ephemerides published in Kleine Planeten by the Berlin Rechen-institut on the basis of these elements and tables, as is indicated by comparison with approximate photographic positions at Königstuhl[7] in 1918 and at Algiers in 1921.[8] Further corrections of the elements should be undertaken only on the basis of perturbations by Jupiter of the second order, and of perturbations by other major planets.

For the group of minor planets having a mean motion of about $750''$ (the Minerva group), D. T. Wilson[9] has computed tables after the method of Bohlin. No application of this theory seems to have been made to (93).

REFERENCES

[1] A. N. vol. 70, p. 45.
[2] A. N. vol. 70, p. 205.
[3] Memoirs of the National Academy of Sciences. vol. viii, third memoir. Washington, 1899.
[4] A. N. vol. 158, p. 95.
[5] A. N. vol. 158, p. 175; A. N. vol. 160, p. 352.
[6] Memoirs of the National Academy of Sciences. vol. x, seventh memoir. Washington, 1910.
[7] Eph. Z, A. N. 1918/558.
[8] C. O. M. 1921/171.
[9] Astronomiska Iakttagelser och Undersökningar. vol. 10, No. 1, Stockholm.

TABLE 7.—Elements—(93) Minerva

Letter	Epoch	M	π	Ω	i
		o / "	o / "	o / "	o / "
A........	1867 Oct. 2.0....Berlin......	66 47 58.8	276 39 54.8	5 2 28.0	8 35 34.9
B........	1867 Oct. 2.0....Berlin......	67 1 59.1	275 38 16.3	5 4 11.4	8 36 31.8
C........	1870 May 1.0....Berlin......	270 51 42.1	275 2 55.0	5 4 16.2	8 36 17.6
D........	1870 May 1.0....Berlin......	270 53 55.4	275 0 36.8	5 4 26.8	8 36 34.6
E........	1872 Nov. 6.0....Berlin......	109 32 42.8	274 43 34.4	5 3 40.3	8 36 34.3
F........	1872 Nov. 6.0....Berlin......	109 32 48.4	274 43 34.4	5 3 40.3	8 36 34.3
G........	1879 Feb. 3.0....Berlin......	241 7 28.6	274 42 35.8	5 9 11.9	8 36 3.2
H........	1911 Jan. 31.5...Berlin.....	236 37 30	277 36 31.2	5 4 31.2	8 35 28.0
I.........	1872 Nov. 2.0....Greenwich..	108 37 48.4	274 47 41.4	5 5 25.0	8 36 21.6
J.........	1872 Nov. 2.0....Greenwich..	108 28 35.7	274 49 19.2	5 5 17.5	8 36 23.6
K........	1875 Jan. 0.0....Greenwich..	278 32 8	274 51 41	5 7 8	8 36 20

Letter	Epoch	φ	μ	Equinox	Computer
		o / "	"		
A......	1867 Oct. 2.0....Berlin............	7 39 29.5	776.43667	1867.0	Lehmann
B......	1867 Oct. 2.0....Berlin............	8 3 47.9	775.5500	1870.0	Lehmann
C......	1870 May 1.0....Berlin............	8 3 55.5	776.41806	1870.0	Lehmann
D......	1870 May 1.0....Berlin............	8 4 38.9	776.51030	1870.0	Lehmann
E......	1872 Nov. 6.0....Berlin............	8 4 43.5	776.47953	1870.0	Lehmann
F......	1872 Nov. 6.0....Berlin............	8 4 45.1	776.49465	1870.0	Lehmann
G......	1879 Feb. 3.0....Berlin............	7 59 4.8	775.63887	1880.0	Lehmann
H......	1911 Jan. 31.5...Berlin............	8 15 30	775.6316	1910.0	Berberich
I......	1872 Nov. 2.0....Greenwich........	8 5 0.5	776.51130	1872.0	Eichelberger
J......	1872 Nov. 2.0....Greenwich........	8 4 52.4	775.920408	1872.0	Eichelberger
K.....	1875 Jan. 0.0....Greenwich........	8 4 54	775.9214	1875.0	Leuschner

(94) AURORA

Discovered by Watson[1], 1867, September 6, at Ann Arbor.

Preliminary elements were computed by Tietjen[2][3], the second set given below as Elements A.

With Elements A, H. Leppig[4] formed eight normal places over 162 days, and determined Elements B. An accurate ephemeris including special perturbations by Encke's method was computed for 1868, December 15 to 1869, January 31. Observations by Vogel, January 15, 17, 18, 1869, gave residuals, Δa —21ˢ.6, $\Delta \delta$ +53″.5.

Thereafter, the B. J. gives elements by Leppig from 1872 to 1915, (but with a misprint of 4° in ω from 1887 to 1897[5]). Bauschinger gives Leppig's Elements C in the "Tabellen"[6] for the epoch 1883, July 12.0. The various elements in the B. J. to 1915 are probably brought up from Leppig's Elements B, with special perturbations. The character of the perturbations is not given. Nor is any reference made to arbitrary corrections.

In Kleine Planeten, 1916, the elements are changed by estimating the perturbations 1883–1910 and roughly determining, M and μ, Elements D.

In Kleine Planeten, 1921, the last elements are again corrected by the computation of Jupiter's perturbations and a representation of observed positions, 1884-1918, within ±1ᵐ.5. Osculating elements, 1921, April 24, not available.

From 1884 to 1899 the planet was practically lost, mainly on account of the misprint in ω. Coddington[7] computed a place with the elements of B. J. 1901, and found the planet Δa +5ᵐ.2, $\Delta \delta$ —20′, 1899. For the computation of the perturbations and tables of the Watson asteroids, Leuschner made a collection of Leppig's elements in the B. J. including Elements C and derived average Elements E from those in B. J. 1871, 1873, 1874, 1875, 1877, 1878, 1881, 1898. From these, approximate mean elements were derived. The perturbations were developed and a preliminary correction of the mean motion and mean anomaly from observations in 1867 and 1899 was attempted. A mistake in one of the main terms of the perturbations was discovered and corrected. Nevertheless, large discrepancies between observation and computation remained (—5° in a, 1867). These differences were used for a preliminary correction of the mean motion and of the mean anomaly. With the new value of the mean motion the perturbations were corrected, the residuals re-determined, and further corrections made to the mean motion and to the mean anomaly. A least square solution of 12 places from 1867 to 1899 was then made including the corrected perturbations. The residuals were thereby reduced from $\pm 2°.2$ to a maximum of $\pm 0°.14$. Elements F.

Further correction of the perturbations by means of the new mean motion produced larger residuals. The best representation as above is obtained by the use of the adopted Elements F without further correcting the perturbations.

From the experience of the Recheninstitut and of Leuschner with Leppig's elements, it is evident that before a satisfactory representation for all oppositions, without making arbitrary corrections to the elements, can be obtained, it will be necessary to derive an accurate set of osculating elements by connecting a limited number of oppositions with accurate determination of the perturbations. With an accurate set of osculating elements the perturbations may then be corrected, but further correction of the elements should be made only after higher order perturbations and perturbations by planets other than Jupiter shall have been considered.

REFERENCES

[1] A. N.   vol. 70, p. 79.
[2] A. N.   vol. 70, p. 219.
[3] A. N.   vol. 71, p. 47.
[4] B. J.   1871.   A. N. vol. 72, p. 331.
[5] A. N.   vol. 139, p. 63.
[6] Tabellen zur Geschichte und Statistik der Kleinen Planeten.
[7] A. N.   vol. 153, p. 225.

TABLE 8.—*Elements—(94) Aurora*

Letter	Epoch	M.T.	M	ω	Ω	i
			° $\prime$ $\prime\prime$	° $\prime$ $\prime\prime$	° $\prime$ $\prime\prime$	° $\prime$ $\prime\prime$
A.....	1867 Nov. 28.0..	Berlin......	340 30 39.5	40 50 23.2	4 32 9.3	8 5 27.0
B.....	1870 Jan. 0.0....	Berlin......	115 9 43.74	40 2 43.08	4 34 36.38	8 5 18.49
C.....	1883 July 12.0..	Berlin......	256 3 4.3	45 22 31.8	4 25 0.9	8 4 14.0
D.....	1925 Jan. 0.5....	Greenwich..	20 40 1.2	57 20 9.6	4 22 26.4	8 4 8.4
E.....	1875 Jan. 0.0...	Greenwich..	73 37 45	41 51 21	4 28 46	8 4 54
F.....	1875 Jan. 0.0...	Greenwich..	65 20 12	48 13 54	4 24 21	8 3 51

Letter	Epoch	M.T.	φ	μ	Equinox	Author
			° $\prime$ $\prime\prime$	$\prime\prime$		
A......	1867 Nov. 28.0	Berlin......	5 10 18.6	630.5129	1867.0	Tietjen
B......	1870 Jan. 0.0........	Berlin......	5 6 8.13	631.5264	1870.0	Leppig
C......	1883 July 12.0........	Berlin......	4 44 18.3	630.6584	1900.0	Leppig
D......	1925 Jan. 0.5........	Greenwich..	5 4 58.8	631.800	1925.0	Berberich
E......	1875 Jan. 0.0........	Greenwich..	4 56 22	631.2196	1900.0	Leuschner
F......	1875 Jan. 0.0........	Greenwich..	5 17 16	631.9473	1900.0	Leuschner

(127) JOHANNA

Discovered by Prosper Henry[1] at Paris, 1872, November 5.

In B. J. 1875 a set of Elements A by Baillaud is published, based on observations Nov. 9, 22, 28.

Preliminary Elements B by Renan[2] are based on seven observations for the opposition 1872–73. An ephemeris for 1874 April and May is also published. An observation on April 17, 1874, shows corrections to the ephemeris Δa $+2^{m}43^{s}$, $\Delta\delta$ —16′.

An improvement on Elements B is made by Renan[3] on the basis of six normal places (1872–1874). The resulting Elements C represent the normal places within $+7''.8$ and $—9''.1$. Renan states these differences are within the limits of error and Elements C may be considered definitive. An ephemeris is then computed for 1876 September.

Elements D are published by Bauschinger.[4] They are by Maywald and are based on oppositions 1872, 1874, 1876, 1879. The special perturbations of Jupiter and Saturn to 1890 are included.

Renan's elements are published and used by B. J. 1882 to 1887. Beginning with B. J. 1888, Maywald's elements are used. An empirical correction was applied to the mean anomalie in 1913 by Berberich.[5]

A correction to the R. I. ephemeris[6] for 1918, November 23, is Δa $—5^{m}.2$, $\Delta\delta$ $—20'$. An improved ephemeris is published[7] for the period 1920, March 21 to April 21, from elements based on 6 oppositions, 1897–1908. Jupiter's perturbations are included. Representation $\pm 0^{m}.3$. Osculation 1921, July 13. The correction to this ephemeris[6] from an observation[8] February 28, 1920, is Δa $+0^{m}.4$, $\Delta\delta$ $—6'$.

Another ephemeris[9] for 1920 by P. Maitre is based on elements published in Connaissance des Temps for 1915 with the mean anomaly corrected, on the basis of observations in 1917 and 1918, ΔM $—1°.305$.

General perturbations applying Bohlin's method for this planet have been published by D. T. Wilson[10] and similar perturbations with Hansen's method, by M. Viljev.[11]

Olson[12] has published general perturbations of the first order by Jupiter. The basic elements are those of Maywald. Jupiter mass 1:1047.568 (Bessel-Schur). Method that of Hansen. Terms of 6th to 8th order are below 1″. No comparison with observations.

REFERENCES

[1] A. N. vol. 80, p. 239.
[2] A. N. vol. 83, p. 349; C. R. vol. 78, 1874, p. 1219.
[3] C. R. vol. 83, 1876, p. 567.
[4] Veröffentlichungen R. I. No. 16.
[5] B. J. 1915, p. 27.
[6] A. N. vol. 208, p. 14.
[7] B—Z der A. N. No. 11, 1920.
[8] B—Z der A. N. No. 14, 1920.
[9] C. O. M. No. 297.
[10] Ast. Iakttagelser och Undersökningar, Stockholm. vol. 10, No. 1.
[11] Bull. Soc. Astr. Russio. No. 22.
[12] Swed. Akad. Handl., Stockholm, 1895.

TABLE 9.—*Elements—(127) Johanna*

Letter	Epoch	M.T.	M	ω	Ω	i
			° ′ ″	° ′ ″	° ′ ″	° ′ ″
A......	1872 Dec. 18.0..........	Berlin......	293 6 46	90 25 17	31 40 11	8 19 42
B......	1874 Apr. 17.0..........	Berlin......	35 3 42	91 13 48	31 41 41	8 17 28
C......	1876 Sept. 5.5..........	Berlin......	223 47 46	90 50 37	31 46 38	8 16 40
D......	1879 Apr. 4.0..........	Berlin......	67 49 52	89 18 47	31 45 2	8 16 48

Letter	Epoch	M.T.	φ	μ	Equinox	Authority
			° ′ ″	″		
A......	1872 Dec. 18.0............	Berlin......	4 36 31	766.23	1872.0	Baillaud
B......	1874 Apr. 17.0............	Berlin......	3 35 48	776.368	1874.0	Renan
C......	1876 Sept. 5.5............	Berlin......	3 46 51	775.9173	1880.0	Renan
D......	1879 Apr. 4.0............	Berlin......	3 51 17	775.7686	1880.0	Maywald

(128) NEMESIS

Discovered by Watson[1] at Ann Arbor on November 25, 1872, and also by Borrelly[2] at Marseilles on December 4, 1872.

Preliminary Elements A were computed by Bossert,[3] based on observations 1872, November 25, December 7, and December 22. He also published a short search ephemeris.

Preliminary Elements B based on observations covering the first five months were published[4] by Leo de Ball. He then forms six normal places from the same series of observations and from these determines Elements C. These elements represent the normal places within —2″.8 and +3″.2. An ephemeris for 1874 is given in the same reference, page 374.

Preliminary Elements D, based on observations 1872, November 25, December 4, and December 12, are published by H. Richter.[5]

Elements E, based on eight normal places from the first two oppositions (1872–1874) were computed by Ball.[6] The normal places are represented in the plane between +2″.13 and —2″.54 and perpendicular to the plane between +2″.7 and —2″.3. The special perturbations of Jupiter and Saturn were taken into consideration.

Further Elements F for a new epoch and mean equinox, in which the special perturbations due to Jupiter and Saturn have been taken into account, have been published by Ball[6] and also an ephemeris for 1875.

The maximum correction to the ephemeris[7] computed from Ball's elements, for July 1880, is Δa +2ˢ.14 and $\Delta \delta$ +12″.0.

Elements G by Palisa are published and used in B. J. 1883, to B. J. 1893. They are based on 5 oppositions 1872–79 and include the perturbations by Jupiter and Saturn.

Ball's elements are again published and used in B. J. 1894 (see Elements H) to B. J. 1913.

Empirical corrections[8] were applied to Ball's Elements H,[9] by Berberich.

The most extensive investigation on this planet is by Leuschner.[10] The Elements I are based on observations extending from 1872 to 1899 and include the general perturbations due to Jupiter of the first order. These elements and perturbations are used in the B. J. 1915 and to date.

The correction to the ephemeris[11] based on Elements I on June 16, 1921, was Δa +0ᵐ.6 and $\Delta \delta$ —1′. Corrections to these elements should be applied only on the basis of higher order perturbations by Jupiter and of perturbations by other planets.

REFERENCES

[1] A. N. vol. 80, p. 319.
[2] A. N. vol. 80, p. 297.
[3] A. N. vol. 80, p. 351.
[4] No. 5, Circular der Berliner Sternwarte. A. N. vol. 82, p. 281.
[5] A. N. vol. 82, p. 63.
[6] A. N. vol. 85, p. 331; vol. 86, pp. 29–31.
[7] A. N. vol. 98, p. 55, and vol. 99, p. 251.
[8] A. N. vol. 189, p. 173.
[9] B. J. 1914.
[10] Mem. of the National Academy of Sciences. vol. x, seventh mem., p. 255.
[11] C. O. M. No. 181.

TABLE 10.—*Elements—(128) Nemesis*

Letter	Epoch	M. T.	M	ω	Ω	i
			° ′ ″	° ′ ″	° ′ ″	° ′ ″
A.....	1873 Jan. 1.0	Greenwich	49 56 20	296 1 28	76 35 50	6 18 27
B.....	1873 Feb. 25.5	Berlin....	59 55 44	298 35 26	76 39 55	6 15 9
C.....	1873 Feb. 25.5	Berlin....	59 57 21	298 34 00	76 40 02	6 15 14
D.....	1872 Nov. 25.0	Berlin....	62 39 10	274 9 56	75 12 46	7 30 40
E.....	1873 Feb. 25.5	Berlin....	59 52 17	298 41 4	76 37 42	6 15 14
F.....	1875 Apr. 25.0	Berlin....	228 43 54	300 3 32	76 30 40	6 15 31
G.....	1880 July 7.0	Berlin....	279 16 11	300 16 4	76 32 4	6 15 43
H.....	1892 Feb. 15.0	Berlin....	116 23 35	299 30 58	76 36 54	6 15 24
I......	1896 July 3.0	Berlin....	101 41 9*	299 56 32	76 39 30	6 15 18

Letter	Epoch	M. T.	φ	μ	Equinox	Authority
			° ′ ″	″		
A.....	1873 Jan. 1.0...	Greenwich	7 21 59	776.86	1873.0	Bossert
B.....	1873 Feb. 25.5...	Berlin....	7 13 5	778.030	1873.0	De Ball
C.....	1873 Feb. 25.5...	Berlin....	7 12 51	778.1516	1873.0	De Ball
D.....	1872 Nov. 25.0...	Berlin....	7 13 17	764.990	1872.0	Richter
E.....	1873 Feb. 25.5...	Berlin....	7 11 59	778.0333	1870.0	De Ball
F.....	1875 Apr. 25.0...	Berlin....	7 13 20	777.4729	1875.0	De Ball
G.....	1880 July 7.0...	Berlin....	7 21 52	777.4964	1880.0	Palisa
H.....	1892 Feb. 15.0...	Berlin....	7 10 53	777.6921	1890.0	De Ball
I......	1896 July 3.0...	Berlin....	7 16 50	777.8761*	1900.0	Leuschner

*Mean Elements.

(175) ANDROMACHE.

Discovered by J. Watson[1] 1877 October 1, but observed only October 5, 6, 16, 29. These were all the observations available until rediscovery May 19, 1893.

By an unfortunate delay the news did not reach other interested observatories until two months later. The preliminary orbit, Elements A, by Tietjen,[2] was erroneous, partly on account of errors in Watson's ringmicrometer observations.

Watson[3] computed Elements B, and with them perturbations for three years until his death in 1880.

Bidschof[4] made an attempt to improve the orbit, Elements C, but perceived the impossibility of this undertaking.

1893, May 19, Charlios[5] at Nice discovered by photography a planet, 1893 Z, and noticed the close similarity between its orbit and that of (175) Andromache. He referred the case to Berberich[6] who followed it out from a preliminary orbit to the best to be obtained from the data in 1893 May 19-August 1, Elements D, and compared his results with the observations in 1877 and with one in 1892 from a photographic plate taken at Heidelberg.

Berberich[7] next computed the extremely large perturbations by Jupiter (in 1887) and Saturn, and improved the orbit by a solution from four normal places in 1877, 1892, 1893, with an ephemeris for the coming opposition, Elements E—"this planet therefore deserves peculiar attention for it will furnish an excellent means for determining an accurate value of the mass of the planet Jupiter." Cf. (3) Juno, (4) Vesta, (13) Egeria, (24) Themis, (33) Polyhymnia, (447) Valentine.

Berberich[8] improved his elements by the following oppositions and brought them forward with the special perturbations of Jupiter and Saturn. The Elements F, G, H, illustrate the large changes in this case of near commensurability (Hecuba group).

This case among others impelled A. O. Leuschner to undertake the computation of the tables[9] for the Hecuba group after Bohlin's method. The application of these tables to the orbit of (175) Andromache was carried out by Miss S. H. Levy. Her unpublished computation contains the transformation of Berberich's elements to Mean Elements I, tables of the perturbations, determination of constants, the comparison with ten oppositions between 1893 and 1907, and at least squares solution which led to the Mean Elements J. The representation of observations in 1914 was $-0^m.4$ in a and $+1'$ in δ.

The comparison between theory and observation has been discussed by A. O. Leuschner.[10]

REFERENCES

[1] C. R. vol. 85, p. 1006. A. N. vol. 91, p. 127.
[2] Circular zum B. J. Nr. 81.
[3] B. J. 1881, 1882, 1886.
[4] Sitz-Ber. Akad. Wien. Bd. 98.
[5] A. N. vol. 132, p. 367.
[6] A. N. vol. 134, p. 143.
[7] A. J. vol. 14, p. 36.
[8] A. N. vol. 153, p. 59.

[9] National Academy of Sciences, Memoirs, vol. 14, third memoir.
[10] A. O. LEUSCHNER, Comparison of theory with observation for the minor planets (10) Hygiea and (175) Andromache with respect to the perturbations by Jupiter. Proc. N. A. S. Washington. vol. 8, No. 7, p. 170.

TABLE 11.—*Elements—(175) Andromache*

Letter	Date	M. T.	M	ω	Ω	i
			° ′ ″	° ′ ″	° ′ ″	° ′ ″
A......	1877 Oct. 29.5....	Berlin............	45 6 25.6	269 26 21.1	23 32 56.0	3 46 38.8
B......	1878 Dec. 5.0....	Berlin............	105 28 46.7	269 31 45.6	23 34 50.2	3 46 36.7
C......	1889 Apr. 7.5....	Berlin............	317 3 18.5	269 42 7.7	23 43 24.9	3 46 45.8
D......	1893 Aug. 1.5....	Berlin............	297 57 33.9	299 49 1.8	25 27 14.8	3 10 51.4
E......	1894 Aug. 23.0....	Berlin............	3 52 54.5	299 46 4.9	25 27 32.5	3 10 59.4
F......	1877 Oct. 11.0....	Berlin............	32 5 0.1	298 31 16.6	25 36 3.8	3 11 42.8
G......	1900 Sept. 1.0...	Berlin............	16 10 41.5	301 33 8.5	25 23 37.7	3 10 38.9
H......	1920 Jan. 20.....	1925.0 Greenwich..	76 24 14	306 45 58	25 7 12	3 10 37
I......	1877 Oct. 11.0....	Berlin............	3 49 12	304 6 54 ˙	25 36 4	3 11 43
J......	1877 Oct. 11.0....	Berlin............	3 59 32	304 13 9	25 43 27	3 11 29

Letter	Date	M. T.	φ	μ	Equinox	Authority
			° ′ ″	″		
A......	1877 Oct. 29.5....	Berlin............	20 26 45.7	542.173	1877.0	Tietjen
B......	1878 Dec. 5.0....	Berlin............	20 26 12.6	541.779908	1880.0	Watson
C......	1889 Apr. 7.5....	Berlin............	20 15 17.8	544.411	1890.0	Bidschof.
D......	1893 Aug. 1.5....	Berlin............	11 39 45.8	614.943	1890.0	Berberich
E......	1894 Aug. 23.0....	Berlin............	11 36 51.4	614.63354	1890.0	Berberich
F......	1877 Oct. 11.0....	Berlin............	12 8 54.1	617.7375	1890.0	Berberich
G......	1900 Sept. 1.0...	Berlin............	11 7 42.9	612.2868	1900.0	Berberich
H......	1920 Jan. 20.....	1925.0 Greenwich..	10 42 11	607.899	1925.0	Berberich
I......	1877 Oct. 11.0....	Berlin............	12 21 26	619.5629 ˙	1890.0	Berberich
J......	1877 Oct. 11.0....	Berlin............	12 19 34	619.025	1890.0	Miss Levy

(433) EROS, 1898 DQ.

Discovered 1898, August 13, by Witt at Berlin and by Charlois at Nice.

Fayet[1] computed Elements A from three observations on August 15, 26, and September 7. Later he[2] computed Elements B from normal places 1898 August 16.5(7 observations), September 17.5(6 observations), and October 22.4(2 observations). The residuals vary from —0^s.11 to —0^s.36 in a, and +5″.1 to +8″.2 in δ.

Hussey[3] gives Elements C, computed from the mean of two Kiel observations made on the 15th of August, and observations made at Mt. Hamilton on September 6 and 27. Hussey[4] later computed Elements D from observations at Mt. Hamilton on August 15, September 27, and November 11, 1898. The residuals for the observations in 1898 vary from +0^s.04 to —0^s.18 in a, and +2″.2 to +4″.4 in δ and observations to May 4, 1899, are closely represented.

Chandler[5] computed Elements E from observations August 14 to November 16, 1898. Elements F[6] and G[7] are also due to Chandler. Elements F are from eight normal places from August 17.5 to November 26.5. The representation is within the errors of observation.

In a later article he gives the following residuals for earlier observations found on plates taken in 1896 at Arequipa:

1896	Δa	$\Delta \delta$
April 6	—0^m 36^s.0	—4.9
June 5	—1 16.7	—5.8

Elements G[7] are the preceding ones with corrections applied so as to fit the observations made at Arequipa 1896. Representation[8] for 1893, 1894, 1896, gives maximum residual of +7^s.6 in a and —1′.0 in δ.

The observations from 1893 to February 16, 1894, were found by Pickering and Mrs. Fleming on plates taken at Cambridge. Perturbations were not considered in applying the corrections to Elements F to obtain Elements G.

Elements H are due to Berberich[9] and are given under the title of "First elements," and are based upon observations made at Urania (Berlin) on the 14, 23, and 31 of August. Berberich gives forty-three sets of residuals covering the period August 13 to August 31, Δa being greater than 0^s.40 but three times, and $\Delta \delta$ being larger than 10″.0 but twice.

Elements I[10] J[11] K[12] and L[13] are due to Russell. Elements I are computed from three normal places obtained by comparison of observations in the A. N. and A. J. with places computed from Berberich's Elements H.[9] (1898, August 18.5, 34 observations, August 26.5, 16 observations, and September 9.5, 26 observations, heliocentric arc about 8°.) The set J[11] is based upon nine normal places, although Dr. Chandler's value of the mean motion was taken (Elements G) and the other elements determined by varying the ratio of the extreme geocentric distances. The normal places are well represented.

Elements K, Russell,[12] are the same as J[11] except for changes in M, ω, and φ based upon observations in 1899.0 at the Chamberlin and Lick Observatories. The representation August 17, 1898, to May 20, 1899, is satisfactory except for the normal place of November 11.5 for which Δa —^{s}.28, $\Delta \delta$ —3″.

Elements L are merely the preceding ones brought up to the epoch 1900.0 and mean equinox of 1900.0. In his article [18] Russell develops the general perturbations of the major axis of Eros by the action of Mars. He does this by Le Verrier's method of interpolation. Russell finds eight terms of the general perturbations of the mean longitude larger than 1″.50. The largest is 35″ with a period of about 1000 years. The greatest displacement due to the first 7 terms will be +38″ in 1927 and —53″ in 1959 in mean longitude, and "will eventually lead to a valuable determination of the mass of Mars." Russell then gives tables of the perturbative function, perturbations of log a and perturbations of the mean longitude.

Elements M [14] were developed by Robbins from Elements G by applying special perturbations of Venus, Earth, Mars, Jupiter, and Saturn by the method of the variation of constants. (Nautical Almanac 1837, appendix.)

Elements N[15] are due to H. Osten, who has computed eight normal places based upon Elements P, with special perturbations of Venus, Earth, Mars, Jupiter, and Saturn according to Encke's method.

Millosevich has produced numerous sets of elements. Elements O[16] were computed from observations made during the interval August 14 to September 21, 1898. An observation by Millosevich October 8, gives Δa —1^s.93, $\Delta \delta$ +7″.5.

Elements P[17] were computed from a normal place of date August 14.5 and Millosevich's observations on September 21 and October 24, 1898. Millosevich states that Berberich's ephemeris requires a correction of +131^s in a and +5′.5 in δ on December 23, 1898.

Elements Q[18] are from photographic observations at Greenwich by the variation of the distances. Later [19] he stated that there is an error of about 2^s in his ephemeris after five months.

Elements R[19] are based upon four normal places and are Elements P improved by the method of variation of the distances. They represent 17 normal places from 1000 observations in 1898–1899 perfectly.

Elements S[20] are based upon 17 normal places made from 999 observations in α and 992 observations in δ in the years 1898–1899, and are derived from Elements R brought forward with the perturbations of Venus, Earth, Mars, Saturn and Jupiter for 20-day intervals.

Elements T[21] U[22] V[23] W[24] X[25] Y[26] are improvements of preceding elements, as are Elements Z[27] AA[27] AB[27] AC[28] AD[29] and AE[30]. Elements AE are Elements AC with special perturbations for 20-day intervals for the period 1901, March 20, to 1903, June 8.0 applied. These perturbations were computed by Wedemeyer, those of Venus, Earth, Mars, Jupiter and Saturn being considered.

Elements AF were found only in the B. J. for 1907. They are probably Elements AE with the epoch changed and perturbations applied. Dziewulski has published "Sekulare Marsstörungen in der Bewegung des Eros.," Bull. de l'Acad. des Science de Cracovie 1905. Not available here. Some mistakes are corrected in A. N. vol. 175, p. 171. In A. N. vol. 175, p. 17, Merfield gives the secular perturbation of Eros due to all major planets. Gauss' method by Hill. Elements from Hill's memoir, Ast. Papers of the Amer. Ephm. Vol. IV and Elements W. The secular perturbations by Jupiter, the Earth and Venus are the largest.

Elements AG are by G. Witt.[31] They are based upon observations from 1893 to 1903, the perturbations of Venus, Earth, Mars, Jupiter and Saturn being included. The perturbations were calculated by the method of variation of constants. For Mars, Jupiter and Saturn the perturbations were calculated for 20-day intervals and for Venus and the Earth at 10-day intervals. With these elements he calculates the perturbations from 1903 to the beginning of 1908, and includes these in an ephemeris for 1905 for dates from July 17 to August 22 (B. J. 1907, p. 476.).

Elements AH[32] AI[33] AJ[34] are preceding elements with change of osculation. Elements AI are also found in the Connaissance des Temps for 1915, but with the equinox changed in 1920.0, Greenwich M. T.

In an article "Beiträge zur Theorie der Bewegung des Planeten 433 Eros," E. Noteboom[35] uses Elements AL to compute the general perturbations of Mercury, Uranus, and Nepture. A recurring run in the residuals is not due to these planets.

Noteboom then uses the twenty normal places of Witt[36] (which lie between dates 1893 October 31 and 1907 October 8) and forms four more normal places, one in 1910, one in 1912, and two in 1914. He then gets Elements AM out of a least square solution by a correction of Elements AL. He gives the mass of Earth and Moon as $1/328370\pm102$ and $\pi=8''.799$.

No authority is available for Elements AK.[37]

REFERENCES

[1] A. N. vol. 147, p. 335.
[2] A. N. vol. 148, p. 27. C. R. 127, p. 806.
[3] A. J. vol. 19, p. 120.
[4] A. J. vol. 20, p. 61. A. N. vol. 148, p. 143. A. J. B. 1899, p. 132.
[5] A. J. vol. 19, p. 148.
[6] A. J. vol. 19, p. 155.
[7] A. J. vol. 19, p. 160. A. J. B. 1899, p. 132.
[8] A. N. vol. 148, p. 189.
[9] A. N. vol. 147, p. 221.
[10] A. J. vol. 19, p. 147.
[11] A. J. vol. 20, p. 8. A. J. B. 1899, p. 132.
[12] A. J. vol. 20, p. 134. A. J. B. 1899, p. 132.
[13] A. J. vol. 21, p. 25.
[14] M. N. vol. 60, p. 614. A. J. B. 1900, p. 158.
[15] A. N. vol. 150, p. 362. A. J. B. 1899, p. 132.
[16] A. N. vol. 147, p. 363.
[17] A. N. vol. 147, p. 397. B. J. 1901.
[18] A. N. vol. 148, p. 271. A. J. B. 1899, p. 132.
[19] A. N. vol. 151, p. 130.
[20] A. N. vol. 151, p. 137. A. J. B. 1899, p. 132. B. J. 1902.
[21] A. N. vol. 153, p. 25.
[22] A. N. vol. 153, p. 25. A. J. B. 1900, p. 158.
[23] A. N. vol. 153, p. 217.
[24] A. N. vol. 153, p. 217. A. J. B. 1900, p. 158.
[25] A. N. vol. 154, p. 142.
[26] A. N. vol. 154, p. 142. B. J. 1903.
[27] A. N. vol. 156, p. 327.
[28] A. N. vol. 156, p. 328. B. J. 1904. A. J. B. 1901, p. 184.
[29] A. N. vol. 155, p. 25. A. J. B. 1901.
[30] B. J. 1905, p. 534 and p. 428.
[31] G. Witt; Untersuchung über die Bewegung des Planeten (433) Eros, Berlin, 1905, Druck der Norddeutschen Buchdruckerei. A. N. vol. 176, p. 211-213.
[32] B. J. 1908, p. 496
[33] B. J. 1909.
[34] B. J. 1916.
[35] A. N. vol. 214, p. 153.
[36] "Uber die Notwendigkeit einer Verbesserung der Masse des Systems Erde-Mond," by G. Witt. Vierteljahrsschrift der A. G. 1908, p. 295.
[37] B. J. 1918, p. (12).

TABLE 12.—Elements—(433) Eros

Letter	Epoch	Mean Time	M	ω	Ω	i	φ	μ	log a	Authority	Equinox
A	1898 Aug. 15.5.	Paris	221 25 0	172 7 7	302 25 38	9 57 35	12 12 14	2018.45	0.163326	G. Fayet	1898.0
B	1898 Aug. 16.5.	Paris	213 1 42.4	177 46 7.5	303 30 19.8	10 49 36.7	12 51 53.9	2016.11	0.163662	G. Fayet	1898.0
C	1898 Aug. 31.5.	Greenwich	222 51 53.3	176 52 17.6	303 23 45.2	10 44 41.3	12 49 40.7	2013.491	0.164038	W. Hussey	1898.0
D	1898 Aug. 31.5.	Greenwich	221 33 29.10	177 41 21.00	303 36 19.2	10 49 33.5	12 52 16.1	2015.775	0.1637097	W. Hussey	1898.0
E	1898 Aug. 31.5.	Greenwich	219 59 23.8	178 40 7.3	303 36 1.1	10 50 31.2	12 51 1.6	2023.656	0.162580	S.C.Chandler	1898.0
F	1898 Aug. 31.5.	Greenwich	221 46 45.6	173 33 32.5	303 29 45.4	10 49 27.2	12 52 32.8	2014.6326	0.1638739	S.C.Chandler	1898.0
G	1898 Aug. 31.5.	Greenwich	221 35 45.6	177 37 56.0	303 31 57.1	10 50 11.8	12 52 9.8	2015.2326	0.1637876	S.C.Chandler	1898.0
H	1898 Aug. 31.5.	Berlin	220 14 3.7	178 28 26.2	303 48 53.0	11 6 57.1	13 13 3.8	2010.131	0.164521	Berberich	1898.0
I	1898 Aug. 31.5.	Berlin	224 33 12.3	175 47 50.1	303 20 20.3	10 45 1.8	12 55 13.6	2003.86	0.1654245	Russell	1898.0
J	1898 Aug. 31.5.	Greenwich	221 38 37.8	177 38 15.2	303 29 57.3	10 49 31.0	12 52 27.9	2015.2326	0.1637876	Russell	1898.0
K	1898 Aug. 31.5.	Berlin	221 37 2.0	177 39 10.6	303 29 57.3	10 49 31.0	12 52 14.2	2015.2326	0.1637876	Russell	1898.0
L	1900 Jan. 0...	Greenwich	133 57 12.7	177 39 6.8	303 31 41.6	10 49 31.6	12 52 14.2	2015.2326	0.1637876	Russell	1900.0
M	1900 Nov. 1.5.	Greenwich	304 57 42.8	177 38 2.9	303 33 18.5	10 50 16.5	12 52 38.3	2015.2423	0.1637863	F. Robbins	1900.0
N	1898 Oct. 1.0.	Berlin	238 39 44.64	177 39 21.05	303 31 53.37	10 49 33.99	12 52 18.33	2015.34326	......	H. Osten	1900.0
O	1898 Aug. 31.5.	Berlin	222 23 28.7	177 9 34.8	303 24 53.1	10 45 18.1	12 49 5.4	2015.119	0.163804	Millosevich	1898.0
P	1898 Aug. 31.5.	Berlin	222 16 27.1	177 13 32.6	303 27 48.3	10 48 33.5	12 52 48.3	2014.656	0.1642210	Millosevich	1898.0
Q	1898 Aug. 2.5.	Berlin	221 40 29.6	177 36 7.3	303 29 50.9	10 49 33.1	12 52 25.4	2014.656	......	Millosevich	1898.0
R	1898 Aug. 2.5.	Berlin	205 22 27.98	177 38 39.02	303 31 46.59	10 49 35.30	12 52 21.99	2015.16324	0.1637975	Millosevich	1900.0
S	1900 Oct 31.5.	Berlin	304 23 59.7	177 38 41.6	303 30 40.4	10 49 38.9	12 52 48.2	2015.12740	0.1638027	Millosevich	1900.0
T	1898 Aug. 2.5.	Berlin	205 21 13.70	177 39 18.84	303 30 44.14	10 49 29.36	12 52 5.98	2015.33037	......	Millosevich	1900.0
U	1900 Oct. 31.5.	Berlin	304 25 2.56	177 39 21.41	303 30 37.99	10 49 32.98	12 52 32.15	2015.29453	......	Millosevich	1900.0
V	1898 Aug. 2.5.	Berlin	205 21 41.83	177 38 55.23	303 31 56.17	10 49 35.35	12 52 14.44	2015.26908	0.1637824	Millosevich	1900.0
W	1900 Oct. 31.5.	Berlin	304 24 40.34	177 39 57.80	303 30 50.02	10 49 38.97	12 52 40.61	2015.22324	0.1637875	Millosevich	1900.0
X	1898 Aug. 2.5.	Berlin	205 21 42.96	177 39 51.72	303 31 51.72	10 49 35.36	12 52 21.16	2015.27302	0.1637818	Millosevich	1900.0
Y	1900 Oct. 31.5.	Berlin	304 24 44.71	177 39 6.18	303 30 45.57	10 49 38.98	12 52 47.33	2015.23718	......	Millosevich	1900.0
Z	1900 Dec. 10.5.	Berlin	326 48 27.27	177 39 8.65	303 30 42.38	10 49 38.03	12 52 49.93	2015.23858	0.1637867	Millosevich	1900.0
AA	1901 Feb. 8.5.	Berlin	0 23 41.36	177 39 13.04	303 31 33.08	10 49 38.75	12 52 53.47	2015.18174	0.1637850	Millosevich	1900.0
AB	1901 Mar. 20.5.	Berlin	22 47 1.51	177 39 12.65	303 30 28.04	10 49 39.89	12 53 0.33	2015.05612	0.1638129	Millosevich	1900.0
AC	1901 Mar. 20.5.	Berlin	22 46 55.3	177 39 12.4	303 30 31.6	10 49 39.54	12 53 1.03	2015.04007	0.1638153	Millosevich	1900.0
AD	1903 June 8.0.	Berlin	115 47 54.3	177 43 17.2	303 30 4.2	10 49 40.4	12 53 0.5	2014.98718	0.1638229	Millosevich	1900.0
AE	1905 Aug. 6.0.	Berlin	197 55 55.8	177 24 21.9	303 37 35.5	10 49 41.9	12 53 21.9	2014.9884	0.1637906	Millosevich	1900.0
AF	1898 Aug. 2.0.	Berlin	205 4 51.833	177 39 11.579	303 31 47.859	10 49 35.040	12 52 24.153	2015.275469	0.1638232	G. Witt	1910.0
AG	1905 Aug. 6.0.	Berlin	197 55 57.2	177 44 21.4	303 37 36.8	10 49 39.0	12 52 54.1	2014.9849	0.1638127	G. Witt	1910.0
AH	1907 Oct. 15.0.	Berlin	285 40 28.0	177 46 3.8	303 37 3.5	10 49 41.2	12 52 58.8	2015.0581	0.1638457	G. Witt	1910.0
AJ	1914 Sept. 28.0.	Berlin	267 11 1.2	177 50 22.3	303 35 8.6	10 49 39.6	12 53 0.5	2014.8293	......	G. Witt	1910.0
AK	1925 Jan. 0.5.	Greenwich	204 35 6.0	177 49 55.2	303 48 10.8	10 49 44.4	12 52 58.8	2014.829	......	G. Witt	1925.0
AL	1898 Aug. 2.0.	Berlin	205 4 51.547	177 39 11.547	303 31 47.859	10 49 35.237	12 52 24.338	2015.275271	......	G. Witt	1900.0
AM	1898 Aug. 2.0.	Berlin	205 4 51.429	177 39 11.419	303 31 48.405	10 49 35.138	12 52 24.579	2015.274706	......	E. Noteboom	1900.0

(447) VALENTINE (1899 ES).

Discovered by Wolf and Schwassmann [1] at Heidelberg 1899, October 27.

Preliminary elements were computed by Kreutz [2] based on observations 1899, October 29, November 11 and December 3. Elements A.

Corrections to the ephemerides based on improved elements in 1902 [3] $\Delta a +16^s$, $\Delta \delta -2'$; in 1905 [4] January 13, $\Delta a +36^s$, $\Delta \delta -0'.3$.

A more complete investigation of the orbit was undertaken by Hans Osten. [5] He received from Kreutz five sets of elements, B to F, which refer to different epochs in order to show the effects of the perturbations. Kreutz states that Elements B to E are comparable and are to be preferred to Elements F. Elements B to E were determined from five normal places (1899 to 1904). The residuals for the normal places are

	Ia	Ib	II	III	IV	V
$\Delta a \cos \delta$	$+5\overset{.}{.}1$	$+8\overset{.}{.}7$	$+1\overset{.}{.}2$	$-3\overset{.}{.}2$	$-0\overset{.}{.}7$	$+35\overset{.}{.}5$
$\Delta \delta$	$+1.4$	$+8.1$	-1.1	$+4.0$	-0.2	-3.0

For the purpose of investigating the general perturbations for this planet, Osten makes use of Kreutz' Elements B, Leverrier's elements for Jupiter and Saturn, except for the adoption of Hill's values of the mean motion for each, for Jupiter's mass Newcomb's value and for Saturn's mass Bessel's value. The perturbations of the first order of the masses are determined according to Hansen's method. As a test on his results, he compares observations with theory for seven normal places (1894 to 1906) with the following results:

		$\Delta a \cos \delta$	$\Delta \delta$
1.	1894	$-10\overset{.}{.}73*$	$- 0\overset{.}{.}54*$
2.	1899a	$+ 5.07$	$+ 1.41$
3.	1899b	$+ 8.28$	$+ 8.06$
4.	1901	$+ 0.32$	$- 1.01$
5.	1902	$+ 2.44$	$+ 1.39$
6.	1904	$+33.26$	$+16.12$
7.	1906	$+32.06$	$- 2.52$

On the basis of these residuals the elements were corrected, which resulted in Elements G. The comparison between observation and computation for the normal places gives:

---

* Weight 1/9. Observation uncertain.

		$\Delta\alpha \cos \delta$	$\Delta\delta$	
1.	1894	$+8\overset{.}{.}76$†	$+7\overset{.}{.}22$†	
2.	1899a	-2.81	-2.36	
3.	1899b	$+1.57$	$+4.86$	
4.	1901	-2.51	$+0.73$	
5.	1902	$+1.86$	$+2.00$	‡
6.	1904	-0.68	-0.31	
7.	1906	$+1.97$	-1.05	

The approximate perturbations due to Mars were found to be in-effective since they remained less than 0″.01.

For the purpose of improving Elements G by means of additional observations, Osten[6] first computes special perturbations due to Jupiter and Saturn by the method of variation of elements, and determines new Mean Elements H from the corrections which the observations successively indicate. For the other six major planets the general perturbations are computed after the method in Tisserand, Vol. 1, Chap. 22. All the masses are taken from Bauschinger, "Tafeln z. Theor. Astr." From this combination of special and general perturbations a set of osculating elements for the ten oppositions 1899–1912 results, forming the basis for the formation of normal places. A detailed investigation of the comparison stars and the observations leads to normal places with relative weights and corrections for magnitude equation in a. The equations of conditions are solved under four different assumptions, thus Elements I without and Elements J with magnitude equation. No definite solution is accepted, but the importance of using a observations free from magnitude equation is emphasized.

In the next work Osten[7] proceeds to determine the perturbations of the second order starting with Elements I. Hansen's method is followed throughout (correcting an error in "Auseinandersetzung II, page 98," which acts nearly as a change of the perturbing mass). A capital difficulty is encountered in the near commensurabilities ($+2\epsilon$ $-5\epsilon' +1\epsilon''$) with a period of 8650 years.

Osten proposes to treat such inequalities by expansion into power series in the time and then eliminate one of the anomalies. For (447) the term $3\epsilon -7\epsilon'$ is thereby much enlarged. A comparison is made between the special and general perturbations by Jupiter and Saturn. Some deviations of the order of 1″ are attributed to the perturbations of the third order. Eight tables contain the perturbations.

In the hope of obtaining as accurate results for (447) as for the major planets Osten[8] completes his former theory and gives the last

† Uncertain.
‡ Normal places from Kreutz.

part of the perturbations of the second order and those of the third in longitude and radius vector. The accuracy of 1″ in 100 years is extremely difficult to obtain. A first test is the comparison between special and general perturbations in the longitude and radius vector. He finds small deviations which probably are due to the computation of the special perturbations.

As a further test Osten gives the comparison with 16 normal places 1894–1918. The Jupiter mass is also included as an unknown. The value 1:1047.49 is found. 1:1047.35 according to Newcomb is adopted. Thus Elements K are found. The representation of the normals from micrometer observations is —1″ to +2″ in the plane.

REFERENCES

[1] A. N. vol. 150, p. 431.
[2] A. N. vol. 151, p. 159.
[3] A. N. vol. 158, p. 271.
[4] A. N. vol. 170, p. 195.

[5] Astronomische Abhandlungen No. 15, 1908.
[6] A. N. vol. 194, No. 4639, p. 113.
[7] A. N. vol. 199, p. 393.
[8] A. N. vol. 210, p. 129.

TABLE 13.—*Elements—(447) Valentine (1899 ES)*

Letter	Epoch	M. T.	M.	ω	Ω	i
			o ′ ″	o ′ ″	o ′ ″	o ′ ″
A......	1899 Dec. 3.5.....	Berlin............	4 21 32	319 16 21	72 18 33	4 49 33
B......	1899 Dec. 5.5.....	Berlin............	4 40 42	319 15 3	72 24 5	4 49 4
	Osc. Nov. 5.0.....					
C......	1901 Feb. 8.0.....	Berlin............	87 2 32	318 53 31	72 24 0	4 49 4
D......	1902 Apr. 4.0.....	Berlin............	167 42 14	318 29 20	72 23 32	4 49 4
E......	1904 Oct. 10.0.....	Berlin............	345 41 51	316 22 10	72 19 44	4 49 5
F......	1906 Feb. 2.0.....	Berlin............	79 55 40	313 38 14	72 25 5	4 49 21
G......	1899 Nov. 5.0.....	Berlin............	358 57 25	319 13 45	72 24 10	4 49 4
H......	1899 Nov. 5.0.....	Berlin............	2 42 27	315 27 43	72 25 38	4 49 9
I.......	1899 Nov. 5.0.....	Berlin............	358 52 18	319 13 42	72 24 16	4 49 4
J.......	1899 Nov. 5.0.....	Berlin............	358 52 21	319 13 42	72 24 11	4 49 3
K......	1899 Nov. 5.0.....	Berlin............	358 57 21	319 13 40	72 24 17	4 49 4

Letter	Epoch	M. T.	φ	μ	Equinox	Authority
			o ′ ″	″		
A......	1899 Dec. 3.5.....	Berlin............	2 36 37	687.012	1900.0	H. Kreutz
B......	1899 Dec. 5.5.....	Berlin............	2 34 33	687.5969	1900.0	H. Kreutz
	Osc. Nov. 5.0.....					
C......	1901 Feb. 8.0.....	Berlin............	2 35 5	687.6846	1900.0	H. Kreutz
D......	1902 Apr. 4.0.....	Berlin............	2 36 17	688.1604	1900.0	H. Kreutz
E......	1904 Oct. 10.0.....	Berlin............	2 40 15	686.5435	1900.0	H. Kreutz
F......	1906 Feb. 2.0.....	Berlin............	2 38 34	687.9066	1900.0	H. Kreutz
G......	1899 Nov. 5.0.....	Berlin............	2 34 32	687.3550	1900.0	Osten*
H......	1899 Nov. 5.0.....	Berlin............	2 25 42	687.3504	1900.0	Osten
I.......	1899 Nov. 5.0.....	Berlin............	2 34 32	687.6016	1900.0	Osten
J.......	1899 Nov. 5.0.....	Berlin............	2 34 32	687.6018	1900.0	Osten
K......	1899 Nov. 5.0.....	Berlin............	2 34 32	687.3884	1900.0	Osten

* Mean elements.

(588) ACHILLES, 1906 TG.

Discovered by Wolff[1] at Heidelberg 1906, February 22.

From observations of February 22, and March 5, Berberich[2] computed Elements A for a circular orbit (search ephemeris for April 1906).

The first elliptic Elements B were published by Berberich.[3] They are based on observations 1906, February 22, March 23, and April 22. An ephemeris for May and June, 1906, is also included. He also points out that the aphelion point lies far beyond Jupiter's orbit, and that the present orbit has had its form and position for a long time.

From Elements B, Charlier[4] finds that Achilles is approximately 55° ahead of Jupiter, consequently very close to one of Lagrange's libration points. He also points out that we may have here a case (as he has shown)[5] where the planet does not remain at the apex of the equilateral triangle, as suggested by Lagrange, but oscillates about it with a period of about 148 years.

Comments on the character of the orbit of Achilles have been published by Berberich,[6] Crommelin,[7] Ristenpart,[8] Stroobant.[9]

On the basis of observations extending from 1906, February 22, to May 19, Bidschof[10] has published a set of Elements C and an ephemeris for 1907. A continuation of the ephemeris with corrections to the same was published by Bidschof.[11] An observation 1907, February 12, gives Δa —51^s, and $\Delta \delta$ +6'.7.

An ephemeris for 1909 based on Elements B. J. 1911 with special perturbations due to Jupiter was published by Franz.[12]

For the following years, to 1919, the B. J. (Kleine Planeten) publishes elements by Bidschof brought forward to a new epoch and mean equinox,[13] Elements D.

As an application of Leuschner's[14] satellite method, Einarsson[15] computed a preliminary orbit, (Elements E), based on observations 1907, February 12, April 15, and June 2. Special perturbations, due to Jupiter, computed by Encke's method, were included for the period covering the observations. The maximum residuals, for nine observations of 1907, were $\Delta a \cos \delta$ —3".6, $\Delta \delta$ ±0".7. Elements E were used, without perturbations, to represent an observation 1906, May 19, with the following results: $\Delta a \cos \delta$ +1'01".8, $\Delta \delta$ —39".2.

The most recent work on Achilles was done by Julie M. Vinter-Hansen.[16] With Bidschof's Elements D, the special perturbations, due to Jupiter and Saturn, were computed from 1906 to 1914. All observations of 1906 and 1907, two of 1913, and one of 1914, were then represented and residuals determined. With these as a basis, eleven normal places were formed, weighted according to number of observations in each.

The residuals, (freed from perturbations), for the 1906 and 1907 normal places are small. For normal place X, (1913), $\Delta\alpha \cos\delta$ $+2781''.2$, $\Delta\delta$ $+1685''.0$; for normal place XI, (1914), $\Delta\alpha \cos\delta$ $+1664''.1$, $\Delta\delta+852''.7$. The resulting orbit improvement gave Elements F. The residuals for the normal places from Elements F vary from $+19''.8$ to $-40''.9$ in $\Delta\alpha \cos\delta$, and $-9''.2$ to $+23''8$ in $\Delta\delta$. With Elements F the special perturbations were recomputed by J. Braae. With Elements F and the new perturbations, new residuals for the normal places were determined and the orbit improved on this basis. When the corrections to the elements were put back into the equations of condition, the residuals were of the same order as prior to the solution. Miss Hansen concludes a normal place must be incorrect. The equations of condition were again solved, with the last normal place omitted. The resulting Elements G represent the ten normal places in $\Delta\alpha \cos\delta$ $-1''.5$ to $+5''.8$ and $\Delta\delta$ $-1''.3$ to $+2''.3$.

The normal place XI gives residuals $\Delta\alpha \cos\delta$ $-1817''.2$, $\Delta\delta$ $-1172''.0$, which indicates that it does not belong to Achilles.

The special perturbations due to Jupiter for 1915 to 1920 are published by Miss Hansen.[17] They were computed from Elements G. These elements are published and utilized in Kleine Planeten since 1920. In Ark. för Math. Bd. 4 Nr. 20 Linders developed the approximate theory for planets near the libration points and gives the principal perturbations of the elements of 588. Cf. Heinrich V. J. S. 1913.

REFERENCES

[1] A. N. vol. 170, p. 353.
[2] A. N. vol. 171, p. 11.
[3] A. N. vol. 171, p. 127.
[4] A. N. vol. 171, p. 213.
[5] Meddelanden fran Lunds Observatorium, No. 18.
[6] Nat. Rund. vol. 21, p. 485–486.
[7] Observatory. vol. 29, p. 352–355. Pop. Ast. vol. 14, p. 472–475.
[8] H. u. E. vol. 18, p. 517–521.
[9] Ciel et Terre. vol. 27, p. 161–164.
[10] A. N. vol. 174, p. 45–48.
[11] A. N. vol. 174, p. 175.
[12] A. N. vol. 180, p. 295.
[13] B. J. 1914, p. 30.
[14] L. O. Pub. vol. vii, p. 455.
[15] In Manuscript (Berkeley).
[16] Pub. og mindre Meddelelser fra Köbenhavns Obs. No. 37.
[17] A. N. vol. 208, p. 345.

TABLE 14.—*Elements—(588) Achilles (1906 TG.)*

Letter	Epoch	M. T.	M.	ω	Ω	i
A......	1906 Mar. 5.5...	Berlin......	(u) 186° 15.′9	....	316° 1.′2	11° 43.′2
			° ′ ″	° ′ ″	° ′ ″	° ′ ″
B......	1906 Feb. 22.5....	Berlin......	48 57 24	120 25 50	315 34 7	10 20 53
C......	1906 Feb. 22.5....	Berlin......	43 45 37	{ 129 24 11 / 129 24 10	315 31 7 / 315 31 58	10 16 36 / 10 16 36
D......	1907 Apr. 15.5....	Berlin......	80 18 12	125 37 50	315 36 2	10 18 25
E......	1907 Apr. 15.66...	Greenwich..	80 23 50	125 25 21	315 35 45	10 18 45
F......	1907 May 28.0....	Berlin......	82 54 47	127 7 10	315 34 26	10 17 53
G......	1907 May 28.0....	Berlin......	84 3 2	125 36 22	315 35 59	10 18 14

TABLE 14—*Elements—(588) Achilles (1906 TG.)* — Continued

Letter	Epoch	M. T.	φ	μ	Equinox	Authority
			° ′ ″	′		
A......	1906 Mar. 5.5......	Berlin........	....	312.32	1906.0	Berberich
B......	1906 Feb. 22.5........	Berlin........	9 38 43	295.133	1906.0	Berberich
C......	1906 Feb. 22.5........	Berlin........	8 10 15	294.703	1906.0 1907.0	Bidschof
D......	1907 Apr. 15.5........	Berlin........	8 42 54	295.464	1910.0	Bidschof
E......	1907 Apr. 15.66......	Greenwich....	8 45 49	295.6847	1907.0	Einarsson
F......	1907 May 28.0........	Berlin........	8 25 19	294.71497	1910.0	Miss Hansen
G......	1907 May 28.0........	Berlin........	8 36 13	295.96333	1910.0	Miss Hansen

(617) PATROCLUS, 1906 VY.

Discovered by Kopff[1] at Heidelberg, October 17, 1906.

Preliminary Elements A were computed by Heinrich[2] based on four observations October 21 to December 7.

From Elements A Charlier[3] has noted that the longitude of Patroclus is approximately 60° behind Jupiter as compared with Achilles and Hector which are approximately 60° ahead of Jupiter. From the same elements, Strömgren[3] has made a similar comparison.

A second set of preliminary Elements B were computed by Heinrich[4] by variation of the distances on the basis of five observations from 1906 October 21 to December 7. On page 339 of the same reference the correction is made that Elements B refer to equinox 1906. Elements B are used to compute the special perturbations, by the method of variation of the constants, and a third set of Elements C with an ephemeris for 1907 is published.[5] A continuation of the ephemeris is published on page 251 of the same reference. Observations November 8 and 10, 1907, show the following corrections to the ephemeris: $\Delta a = -34^s$ $\Delta \delta = -0'.8$.

Ephemerides for the oppositions 1908, 1909, 1910, 1912, 1913, by Heinrich[6] are based on Elements C. The correction to the ephemeris in 1909 was $\Delta a -39^s$ and $\Delta \delta +3'.3$.

From 1914 to 1918 the ephemerides based on Elements C are published in Kleine Planeten.

An extensive application of Wilkens' method[7] for planets of Jupiter's group, is made by Drucker.[8] He first takes Heinrich's Elements C as a basis for the computation of the perturbations due to Jupiter and Saturn, by the method of variation of elements. Then corrects Elements C on the basis of twelve normal places (1906 to 1918). See Elements D. The residuals for the twelve normal places, from Elements D plus the special perturbations due to Jupiter and Saturn, vary from $-7''.9$ to $+10''.3$ for $\Delta a \cos \delta$ and $-4''.8$ to $+4''.9$ for $\Delta \delta$.

With Elements D the perturbations due to Jupiter and Saturn are again computed and with the former twelve normal places and a new thirteenth normal place, (1919), the Elements D were corrected to a new set, E. With this new set of Elements E, and the new perturbations, the residuals for the normal places vary from $-3''.4$ to $+3''.5$ for $\Delta a \cos \delta$ and from $-2''.8$ to $+2''.9$ for $\Delta \delta$. He states these Elements E may be considered as definite. In a footnote on page 27 of the same reference, the residuals for an observation 1920, December 16, are $\Delta a \cos \delta +4''.4$ and $\Delta \delta +1''.4$. The ephemeris for 1920 was computed from Elements E, plus special perturbations due to Jupiter and Saturn.

Then he derives corresponding elements in Wilkens' system, Elements F and G. Elements F correspond to Elements D, and Elements G correspond to Elements E. He states that it was only necessary according to Wilkens' method to compute Elements F.

He now computes positions (a and δ), for the thirteen normal places, (1) from Elements E, with special perturbations due to Jupiter; (2), the same places from Elements E without perturbations; (3), the same places from Elements G without perturbations; (4), the same places from Elements E with special perturbations due to Jupiter and Saturn. A comparison is then made between the positions computed by the various systems, (1), (2); (3), (4). The table of differences shows that Saturn's perturbations do not become effective until 1918. It shows, as we should expect, that for the period of osculation, the various systems give the same results. It finally shows that positions computed from (3) come closer to those computed by (1), than the places computed by (2).

Finally he computes the special perturbations due to Jupiter, by the variation of elements, with the aid of Elements G and shows that they are of the same order as those computed with Elements E.

The ephemerides published in Kleine Planeten for 1920 and 1921 are based on Elements E, plus special perturbations due to Saturn and Jupiter.

An ephemeris for the opposition using Elements E, 1918-1919, was computed by Paul Maitre.[9] Ephemerides for oppositions 1919-1920, 1920-1921, and 1922, are also published by Drucker.[10]

REFERENCES ·

[1] A. N. vol. 172, p. 387.
[2] A. N. vol. 175, p. 87.
[3] A. N. vol. 175, p. 89.
[4] A. N. vol. 175, p. 291.
[5] A. N. vol. 176, p. 193.
[6] A. N. vol. 179, p. 223—vol. 180, p. 45—vol. 183, p. 207—vol. 190, p. 395—vol. 194, p. 208.
[7] A. N. vol. 205, No. 4906.
[8] A. N. vol. 214, No. 5114.
[9] Cir. O. M. No. 85, 1918.
[10] B.-Z. der A. N. No. 14, 1919—No. 48, 1920—No. 1, 1922.

TABLE 15.—*Elements (617)—Patroclus (1906 VY)*

Letter	Epoch	M.T.	M	ω	Ω	i
			° ′ ″	° ′ ″	° ′ ″	° ′ ″
A......	1906 Oct. 21.5..........	Berlin......	41 31 55	297 28 37	43 21 39	22 16 47
B......	1906 Nov. 29.0..........	Berlin......	41 27 30	302 11 27	43˙25 32	22 3 33
C......	1907 Dec. 14.0..........	Berlin......	73 1 25	302 25 48	43 28 36	22 3 15
D......	1906 Nov. 29.0..........	Berlin......	42 00 14	−58 26 46	43 27 49	22 7 00
E......	1906 Nov. 29.0..........	Berlin......	41 59 24	−58 25 39	43 27 49	22 6 57
F......	1906 Nov. 29.0..........	Berlin......	42 19 5	−58 45 27	43 27 49	22 7 00
G......	1906 Nov. 29.0..........	Berlin......	42 18 15	−58 44 21	43 27 49	22 6 57

TABLE 15—*Elements—(617) Patroclus (1906 VY.)*—Continued

Letter	Epoch	M.T.	φ	μ	Equinox	Authority
			° ′	′		
A......	1906 Oct. 21.5...........	Berlin......	8 42 41	300.145	1906.0	Heinrich
B......	1906 Nov. 29.0...........	Berlin......	8 16 7	300.659	1906.0	Heinrich
C......	1907 Dec. 14.0...........	Berlin......	8 14 38	300.532	1910.0	Heinrich
D......	1906 Nov. 29.0...........	Berlin......	8 15 33	300.7805	1910.0	Drucker
E......	1906 Nov. 29.0...........	Berlin......	8 15 32	300.7654	1910.0	Drucker
F......	1906 Nov. 29.0...........	Berlin......	8 13 9	301.4459	1910.0	Drucker
G......	1906 Nov. 29.0...........	Berlin......	8 13 9	301.4309	1910.0	Drucker

(624) HECTOR, 1907 XM.

Discovered by Kopff[1] at Heidelberg on February 10, 1907.

A preliminary orbit, Elements A, was computed by Strömgren[2] based on observations from February 10 to April 16. These eléments gave the following residuals:

1907	$\Delta\lambda$	$\Delta\beta$
Feb. 10	—1″.0	+2″.2
April 19	+2 .8	+8 .3

He reports that Hector is another planet with mean motion nearly equal to that of Jupiter. Since the perturbations of Jupiter are small and the perturbations of the other planets will be ineffective for a long time, this planet will remain in the neighbourhood of the libration point for a long time.

An observation by Palisa February 29, 1908, gives a correction to the ephemeris[3] based on Elements A as follows: $\Delta a = -37^s \; \Delta\delta = +6'.3$. An ephemeris is published by Strömgren[4] for the opposition of 1909 based on Elements A.

In preparation for the ephemeris of 1911, Strömgren,[5] assisted by J. Fischer-Petersen, computes a new set of Elements B, based on nine normal places (oppositions 1907, 1908, 1909), taking into account the perturbations due to Jupiter and Saturn. These elements represent the normal places, $\Delta a\cos\delta$ between +0′.30 and —.17, and $\Delta\delta$ between —0′.04 and +0′.16.

The planet was next observed in July 1911. The comparison between observation and ephemeris (from Elements B) was unsatisfactory. Strömgren, assisted by Ruben Andersen,[6] again investigated the orbit based on observations from 1907 to 1911. For this purpose Elements A were used to compute the residuals for five observations, July 4 to 16, 1911. A tenth normal place was formed from these five places with the following residuals $\Delta a\cos\delta = -52' 23''.9 \quad \Delta\delta = -11' 36''.6$. A weight of five (number of observations) was given to this normal place. Combining this normal place with the solution that led to Elements B, a new set of elements was derived, Elements C. With these elements the special perturbations, due to Jupiter and Saturn, were computed by the method of Encke, for 1906 to 1912. The observations for 1907 and 1908 were then computed without taking into account the perturbations (reported as being small) and the observations of 1909 and 1911 were represented with perturbations. From these representations ten new normal places were formed and new Elements D were obtained from a least square solution. This set represents the normal places, $\Delta a \cos \delta$ between $-7''.1$ and $+3''.8$, and $\Delta \delta$ between $-1''.6$ and $+3''.0$.

With Elements D and special perturbations, the maximum residuals for opposition of 1912 were $\Delta a \cos \delta = +2^s.12$ and $\Delta \delta = +26''.6$. Strömgren, assisted by Julie M. Vinter Hansen,[7] again investigated the orbit of Hektor on the basis of thirteen normal places (1907 to 1912). A least square solution led to Elements E. At the conclusion of this work,[7] osculating elements were computed for the epoch of each year, 1907 to 1913, based on Elements E and the special perturbations published in the same reference. The ephemeris for 1913 was published in A. N., vol. 198, p. 367.

In A. N., vol. 200, pp. 79–82, Strömgren publishes the comparisons between the observations and the ephemeris, which was based on Elements E and the perturbations due to Jupiter and Saturn, for 1913 and 1914. The maximum residual for 1913 for $\Delta a \cos \delta = +0^s.53$ and for $\Delta \delta = +7''.1$. For 1914 the maximum residual for $\Delta a \cos \delta = +0^s.51$ and $\Delta \delta = +11''.6$. Strömgren regards these residuals as satisfactory.

As an application of Leuschner's satellite method, S. Einarsson[8] computed a preliminary orbit of Hector based on observations 1907, February 10, March 11, and April 16, Elements F. These elements represented an observation of May 2, 1908, as follows: $\Delta a \cos \delta = -4' 32''.4$ $\Delta \delta = +3' 31''.9$. With Elements F, the special perturbations by Encke's method were computed for the 1907 opposition. New elements were then computed from observations 1907, February 10, March 11, and a normal place from observations April 12, 16, and 19, by a differential correction of Elements F. These new Elements G represented the opposition of 1908, May 2, as follows: $\Delta a + 30''$, $\Delta \delta +$ 95'', and for 1909, April 17, $\Delta a + 6'.1$; $\Delta \delta - 6'.7$. The application of Leuschner's method thus yielded far better results than the ordinary method followed by Strömgren: 1908, $\Delta a = -9' 15''$ $\Delta \delta = +6' 18''$. No further work was done on this planet by Einarsson since Strömgren and his colleagues had already made extensive investigations.

An ephemeris for 1915 is published by J. Fischer-Petersen,[9] using Elements E and taking into account perturbations due to Jupiter and Saturn.

The elements and ephemeris for 1916[10] are based on Strömgren's Elements E with special perturbations by Jupiter and Saturn brought forward.

Elements and ephemerides are given in Kleine Planeten for each year to 1921. No further comment is published regarding the elements.

An ephemeris is published by M. Henri Blondel[11] for the opposition of 1921 which has been corrected on the basis of an observation at

Algiers[12] on May 6, 1921. This observation gave a correction to the ephemeris published in Kleine Planeten of $+7^m.0$ in right ascension and $-80'$ in declination.

In A. N., vol. 215, p. 249, A. Wilkens has published an article, "Uber die Säkularen Veränderungen der Grossen Achsen der Bahnen der Planeten der Jupiter Gruppe."

In A. N., vol. 175, p. 89, Charlier gives a brief discussion on the orbits of the Trojan group, regarding their motion about the libration points.

In A. N., vol. 206, p. 235, A. Koref outlines his investigation regarding the motion of Hector. His preliminary work is based on eighteen normal places (1907 to 1914). The investigation will be completed when observations of 1918 and 1919 are available.

REFERENCES

[1] A. N. vol. 174, p. 63.
[2] A. N. vol. 175, p. 14.
[3] A. N. vol. 177, p. 123.
[4] A. N. vol. 180, p. 327.
[5] Publikationer og mindre Meddelelser fra Köbenhavns Observatorium No. 6. A. N. vol. 188, p. 395.
[6] Publikationer og mindre Meddelelser fra Köbenhavns Observatorium No. 8.
[7] Publikationer og mindre Meddelelser fra Köbenhavns Observatorium No. 12.
[8] In manuscript (Berkeley).
[9] A. N. vol. 201, p. 335. B. A. J. 1917.
[10] Eph. der Kleinen Planeten, 1916, p. 17, p. 77, p. 93, p. 97.
[11] Cir. O. M. No. 480 (1921).
[12] Cir. O. M. No. 171, second series.

TABLE 16.—*Elements—(624) Hector (1907 XM)*

Letter	Epoch	M.T.	M	ω	Ω	i
			o ′ ″	o ′ ″	o ′ ″	o ′ ″
A......	1907 Feb. 10.0.........	Berlin......	335 47 12	183 51 52	341 58 25	18 7 17
B......	1907 Feb. 10.0.........	Berlin......	343 51 43	175 6 42	341 58 57	18 8 34
C......	1907 Feb. 10.0.........	Berlin......	345 38 38	173 5 26	341 59 47	18 9 13
D......	1907 Feb 10.0.........	Berlin......	343 40 12	175 19 0	341 59 18	18 8 50
E......	1907 Feb. 10.0.........	Berlin......	343 48 55	175 9 30	341 59 15	18 8 45
F......	1907 Mar. 11.36........	Greenwich..	341 56 43	179 46 25	341 57 32	18 8 05
G......	1907 Mar. 11.36........	Greenwich..	348 27 23	172 44 30	341 57 14	18 8 19

Letter	Epoch	M.T.	φ	μ	Equinox	Authority
			o ′ ″			
A......	1907 Feb. 10.0.........	Berlin......	2 8 24	292.5842	1907.0	Strömgren
B......	1907 Feb. 10 0.........	Berlin......	1 56 46	293.1585	1910.0	Strömgren
C......	1907 Feb. 10.0.........	Berlin... ..	1 43 45	295.3661	1910.0	Strömgren
D......	1907 Feb. 10.0.........	Berlin......	1 56 52	293.1072	1910.0	Strömgren
E......	1907 Feb. 10.0.........	Berlin......	1 56 29	293.1782	1910.0	Strömgren
F......	1907 Mar. 11.36.........	Greenwich..	2 03 07	292.7487	1910.0	Einarsson
G......	1907 Mar. 11.36.........	Greenwich..	1 55 32	293.1164	1910.0	Einarsson

(659) NESTOR, 1908 CS.

Discovered by Wolf[1] at Heidelberg, 1908, March 23.

From observations of 1908, March 25 and May 2, Ebell[2] computed a circular orbit. From these Elements A it is evident that the planet belongs to the Trojan group, Achilles type, and he notes the planet's position is near a libration point (60° ahead of Jupiter).

Preliminary elliptic Elements B, with an ephemeris for 1908, were computed by Ebell[3] from observations 1908, March 23, April 26, and May 19. With Elements B Ebell[4] publishes an ephemeris for 1909.

The improvement of Elements B was undertaken by Anderson.[5] Special perturbations due to Jupiter and Saturn were computed with Elements B. Six normal places were formed from eleven observations extending from 1908, March 23 to 1912, September 9.

The coefficients for the differential equations were computed by the method given in Oppolzer.[6] The solution of the equations gave abnormal corrections when both or either normal places IV and V were used. (IV and V are single observations). His final solution was based on the first four normal places, (1908 to 1909). The resulting Elements C were used to compute the special perturbations due to Jupiter and Saturn and an ephemeris for 1913.

Wolf reports [7] that Nestor cannot be found at ephemeris position.

An ephemeris for 1914 based on Elements C, plus perturbations, (disturbing planets not stated), was published by Andersen.[8] He states no other observations for this planet are available. Reported observations since 1909 do not appear to belong to Nestor.[9]

An ephemeris for 1915 based on Elements C by Anderson, was published by Strömgren.[10] He states the correction to ephemeris, based on an observation of 1914, December 20, is $\Delta\alpha$ +62^s, $\Delta\delta$ —7'.6.

Another attempt to improve the orbit of Nestor was made by Andersen[11] in 1917. In this attempt six normal places were formed from observations in 1908, 1909, and 1917. The perturbations due to Jupiter and Saturn were computed from Elements C, and the normal places represented from the same elements. The starting residuals for observation, 1908 and 1909, were small, (maximum —2".4); for the normal place V, (two observations of 1917), $\Delta\alpha$ cos δ —2008".8, $\Delta\delta$ +458".4; for the normal place VI, (two observations of 1917), $\Delta\alpha$ —1934".7, $\Delta\delta$ +392".2. The differential corrections were computed as in the earlier work.[6] For resulting elements see D. These elements left the following residuals for the V and VI normal places:

	$\Delta\alpha$ cos δ	$\Delta\delta$
V	—4".7	—6".2
VI	+5.0	+7.2

Elements D are sufficiently accurate to show that observations reported for this planet in 1914-1915 do not belong to Nestor.

With Elements D and perturbations computed from Elements C an ephemeris was computed for 1918.

Julie M. Vinter-Hansen has published[12] the special perturbations, (disturbing planets not stated), computed by Pedersen from 1907 to 1918. Also Andersen's Elements D brought up to epoch of 1918, (Elements E), and an ephemeris for 1919.

Seagrave has published[13] an ephemeris for 1920. No reference is made as to what elements were used.

The most recent work on Nestor is by Kristensen.[14] He begins with Andersen's Elements D, and computes the special perturbations due to Jupiter and Saturn and then represents all the observations available from 1908 to 1919. The maximum initial residuals (perturbations included), for the 1919 observations are Δa cos δ —64^s.29, $\Delta\delta$ +494″.8. He then forms eleven normal places based on all the available observations. The maximum initial residuals appear in the IX Normal place, Δa cos δ —963″.0, $\Delta\delta$ +494″.0, (perturbations included). The resulting solution of the eleven equations gives Elements F. These elements represent the normal places satisfactorily. The maximum residuals are Δa cos δ +0^s.16, $\Delta\delta$ +7″.2. He states these Elements F may be considered as a definitive system.

The elements published in Kleine Planeten, 1917 to 1920, are those of Andersen (Elements C), brought up to new epochs.

REFERENCES

[1] A. N. vol. 177, p. 287.
[2] A. N. vol. 177, p. 399.
[3] A. N. vol. 178, p. 71.
[4] A. N. vol. 180, p. 213.
[5] A. N. vol. 195, No. 4678, p. 433.
[6] Oppolzer. vol. ii, p. 390-391.
[7] A. N. vol. 196, p. 14.
[8] A. N. vol. 199, p. 221.

[9] Veröff. R. I. No. 42, remarks 5 and 15.
[10] A. N. vol. 200, p. 56.
[11] A. N. vol. 206, No. 4923, p. 17.
[12] A. N. vol. 208, p. 15-16.
[13] A. J. vol. 32, p. 167.
[14] Pub. og mindre Meddelelser fra Köbenhavns Obs. No. 37.

TABLE 17.—*Elements (659)—Nestor (1908 CS.)*

Letter	Epoch	M. T.	M.	ω	Ω	i
A.........	1908 Mar. 25.5...	Berlin.....	(u) 196° 55′.8	....	350° 55′.7	4° 40′.1
			° ′ ″	° ′ ″	° ′ ″	° ′ ″
B.........	1908 Mar. 23.5...	Berlin.....	240 38 5	327 31 28	349 57 42	4 31 15
C......... Osculation...	1908 Mar. 23.5 ⎱ .. 1908 Apr. 12.0 ⎰	Berlin.....	240 3 56	328 4 54	350 0 1	4 31 31
D........ Osculation...	1908 Mar. 23.5 ⎱ .. 1908 Apr. 12.0 ⎰	Berlin.....	237 29 22	329 41 1	350 2 4	4 31 47
E.........	1918 Mar. 11.0...	Greenwich..	179 57 36	331 58 33	350 5 22	4 31 45
F......... Osculation...	1908 Mar. 23.5 ⎱ .. 1908 Apr. 12.0 ⎰	Berlin.....	238 18 00	329 7 41	350 1 29	4 31 44

TABLE 17.—*Elements (659)—Nestor (1908 CS.)*—Continued

Letter	Epoch	M. T.	φ	μ	Equinox	Authority
			° ′ ″	″		
A.........	1908 Mar. 25.5...	Berlin.....	....	290.67	1908.0	Ebell
B.........	1908 Mar. 23.5...	Berlin.....	6 23 59	300.785	1908.0	Ebell
C........ Osculation...	1908 Mar. 23.5 } .. 1908 Apr. 12.0 }	Berlin.....	6 26 44	301.0002	1910.0	R. Andersen
D........ Osculation...	1908 Mar. 23.5 } .. 1908 Apr. 12.0 }	Berlin.....	6 1 11	301.5134	1910.0	R. Andersen
E........	1918 Mar. 11.0...	Greenwich..	6 2 00	299.803	1919.0	J. M. V. Hansen
F......... Osculation...	1908 Mar. 23.5 } .. 1809 Apr. 12.0 }	Berlin.....	6 7 13	301.2194	1910.0	Kristensen

(716) BERKELEY, 1911 MD.

Discovered by Palisa[1] at Vienna on July 30, 1911.

The first preliminary Elements A were computed by Hopfner[2] and are based on observations 1911, July 30, August 17, and September 3. He also gives an approximate ephemeris for 1911.

Hopfner's Elements A represented the observation of 1906, July 16, for (1906 UN[b]) as follows: $\Delta\alpha$ —1^m.4 $\Delta\delta$ —4'.8. Berberich[3] announces (1906 UN[b]) is identical with (716) Berkeley.

An ephemeris is published by Cohn[4] for 1912. An observation by Palisa[5] for November 4, 1912, gives a correction to the ephemeris, $\Delta\alpha$ —2^m.7 and $\Delta\delta$ —8'.

A new set of Elements B computed by Stracke[6] is based on observations 1911, August 3, 29, and September 23. These elements represent the observation of 1906, July 16, for (1906 UN[b]) as follows: $\Delta\alpha$ +2^m.4 and $\Delta\delta$ +4'.

Stracke's elements are published and used in B. J. 1915 and up to 1919 in Kleine Planeten.

An orbit, Elements C, based on three normal places 1911, July 30, August 22, Septemer 17, was computed by Neubauer[7] using Leuschner's Short Method.[8] A comparison between observation and computation, for observations in 1906, 1912, 1914, of Elements C and Stracke's Elements B, showed that Elements C gave the better representation. Neubauer also computes general perturbations by the Bohlin method,[9] with the tables computed by Wilson[10] for the group 750.[11] More accurate starting elements connecting several oppositions by special perturbations seem to be desirable.

REFERENCES

[1] A. N. 189, p. 109.
[2] A. N. 189, p. 244.
[3] A. N. 189, p. 364.
[4] A. N. 192, p. 426.
[5] A. N. 193, p. 62.
[6] A. N. 192, p. 421–423.
[7] L. O. Bull. No. 301.
[8] L. O. Pub. vol. vii, part viii.

[9] "Formeln and Tafeln zur gruppenweise Berechnung der Allgemeinen Störungen Benachbarter Planeten," Upsala 1896, and "Sur le Dévelopement des Perturbations Planétaires," Stockholm 1902.
[10] Ast. Iakttagelser Och Undersökningar, Band 10, No. 1, Stockholm.

TABLE 18.—Elements (716)—Berkeley (1911 MD)

Letter	Epoch	M.T.	M	ω	Ω	i
			° ′ ″	° ′ ″	° ′ ″	° ′ ″
A......	1911 Aug. 17.5..........	Berlin......	120 23 1	46 3 5	146 51 41	8 37 59
B......	1911 Aug. 18.5..........	Berlin......	118 6 10	48 49 6	146 57 7	8 27 43
C......	1911 Aug. 22.4..........	Berlin......	118 25 50	48 49 15	146 54 49	8 23 38

TABLE 18.—*Elements (716)—Berkeley (1911 MD)*.—Continued

Letter	Epoch	M.T.	φ	μ	Equinox	Authority
			° ′ ″	″		
A......	1911 Aug. 17.5...........	Berlin......	5 28 46	753.940	1911.0	Hopfner
B......	1911 Aug. 18.5...........	Berlin......	5 5 17	754.565	1911.0	Stracke
C......	1911 Aug. 22.4...........	Berlin......	5 23 38	753.7233	1911.0	Neubauer

(718) ERIDA, 1911 MS.

Discovered by Palisa[1] at Vienna, 1911, September 29.

Preliminary Elements A, and an ephemeris based on observations 1911, September 29, October 13, and October 28, are published by Cohn.[2] These elements were used by the B. J. 1915 and 1916.

New Elements B were obtained by Strehlow[3] from observations 1914, February 28, March 18, and March 29. The method of the variation of the distances was utilized so that observations of 1911 and 1914 were well represented.

A correction of $+1''.5$ to μ was applied in order to represent observations of 1904 for (1904, OD) which is supposed to be identical to (718) Erida.

As a test of Leuschner's Short Method for computing orbits,[4] Mundt[5] has published two sets of elements C and D, which are based on the observations used by Cohn.[2]

For the purpose of comparison, ephemeris places were computed from Cohn's Elements A, Strehlow's Elements B, and Mundt's Elements C.

1917 G. M. T.	a	δ	
Oct. 22.5	2^h56^m3	$+17°\ 09'$	Elements C.
Oct. 22.5	2 55.9	$+17\ \ 07$	Elements B.
Oct. 22.5	2 53.5	$+16\ \ 56$	Elements A.
Dec. 1.5	2 25.3	$+15\ \ 39$	Elements C.
Dec. 1.5	2 25.1	$+15\ \ 36$	Elements B.
Dec. 1.5	2 22.9	$+15\ \ 24$	Elements A.

No comparison with observations in 1917 has been made.

Strehlow's Elements B have been published and utilized by B. J. (Kleine Planeten) since 1915. An observation by Palisa[6] on 1919, January 6, gave corrections to the ephemeris as follows: $\Delta a - 1^m.7$ $\Delta\delta - 3'$.

The object of Mundt's work was to test the possibility of deriving by properly chosen methods as satisfactory elements from a single opposition as are ordinarily obtained from at least two oppositions. This result was realized in this case. His work was duplicated by Miss Easton[5] on a slightly different plan of removing the residuals. Elements D.

REFERENCES

[1] A. N. vol. 189, p. 295.
[2] A. N. vol. 192, p. 421–423.
[3] B. J. 1917, p. 36 and 106.
[4] L. O. Pub. vol. vii.
[5] L. O. Bull. No. 302.
[6] E. Z. der A. N. 1919, No. 561.

TABLE 19.—*Elements (718) Erida (1911 MS)*

Letter	Epoch	M.T.	M	ω	Ω	i
			° ′ ″	° ′ ″	° ′ ″	° ′ ″
A......	1911 Sept. 29.5..........	Berlin......	149 0 40	169 56 47	39 22 47	7 3 55
B......	1914 Apr. 1.5..........	Berlin......	320 18 15	168 8 30	39 44 16	6 58 13
C......	1911 Oct. 13.4..........	Berlin......	156 34 10	166 36 12	39 42 41	6 59 5
D......	1911 Oct. 13.4..........	Berlin......	156 03 25	166 55 59	39 43 51	6 58 49

Letter	Epoch	M.T.	φ	μ	Equinox	Authority
			° ′ ″	″		
A......	1911 Sept. 29.5..........	Berlin....	12 5 35	664.65	1911.0	F. Cohn
B......	1914 Apr. 1.5..........	Berlin....	11 28 39	664.412	1910.0	Strehlow
C......	1911 Oct. 13.4..........	Berlin....	11 19 7	663.865	1911.0	C. Mundt
D......;	1911 Oct. 13.4..........	Berlin....	11 20 11	663.769	1911.0	Miss E. J. Easton

(884) PRIAMUS, 1917 CQ.

Discovered by Wolf[1] at Heidelberg on September 22, 1917.

Wilkens outlines[2] his preliminary investigation regarding the motion of this fifth member of the Trojan group. For this purpose he utilizes preliminary Elements A, by Berberich. The libration point is 60° behind Jupiter and the planet oscillates about this point in approximately 150 years. He states that μ varies between 294".27 and 303".99. He also points out that his succeeding work will show that Priamus and Patroclus are diametrically opposite in the small libration ellipse.

Klose applies[3] Wilkens' method[4] for taking into account the principal perturbations of Jupiter, to Priamus. In this method the principal perturbations by Jupiter are accounted for by centering in the Sun the mass of the Sun-Jupiter system. For this study Klose utilizes Berberich's Elements A, bringing them up to mean equinox 1925.0, Elements B, and compares the coordinates computed from Elements B plus special perturbations with those computed from Elements C, which were derived by Wilkens' method. The difference in the representation by the two sets of elements for observations from 1917 October 14 to 1918 December 28, did not exceed 0^s.5 in right ascension and 2" in declination. Klose concludes that Wilkens' method is applicable beyond this short period for immediate Ephemeris purposes.

As an example of his method[5] of integrating the differential equations for the perturbations in the coordinates for planets of the Jupiter group, Wilkens[6] gives a numerical application for Priamus. His results are similar to those of Klose. Klose[7] publishes a further comparison between the usual method of special perturbations due to Jupiter and Wilkens' method. For this purpose he compares results gotten from Elements B and a new set of Elements D, derived by Wilkens' method. He then brings up Elements B with perturbations due to Jupiter up to epoch 1918 (Elements E), and Elements D are brought forward to epoch 1918 by Wilkens' method. An ephemeris is published for the opposition 1919 for which the perturbations of Saturn are also taken into account.

In an article on "Bemerkenswerte Eigenschaften der Bahnen der Planeten der Jupitergruppe," Wilkens[8] points out that the ascending nodes of the six Trojan planets lie with one exception, (Patroclus), in the same quadrant. He also forms a mean value of the ascending nodes from certain planets of the group and compares the individual values with these mean values. He also refers to his article[9] regarding the time of maximum and minimum for the mean motion of Priamus and Patroclus. He finds the maximum of Priamus (1986.8) takes place when the planet is on a line with and between the point of oscillation and the Sun, and the minimum for Patroclus (1985.8) when the planet is on the line with and beyond the point of oscillation and the Sun. This verified his previous conclusion[2] that the two planets are diametrically opposite in their paths of oscillation.

A set of elements and an ephemeris are published in Kleine Planeten for 1920. The elements are probably Elements B, brought up to epoch 1925. An ephemeris is published in Kleine Planeten for 1921.

In A. N. Vol. 215, No. 5147, Wilkens outlines his investigation regarding the secular variations of the major axes of the orbits for the Trojan group.

REFERENCES

[1] A. N. vol. 205, p. 141. M. N. 78, p. 289.
[2] A. N. vol. 207, p. 9.
[3] A. N. vol. 207, p. 183.
[4] A. N. vol. 205, No. 4906.
[5] A. N. vol. 206, No. 4937.
[6] A. N. vol. 208, No. 4984.
[7] A. N. vol. 209, No. 5016.
[8] A. N. vol. 215, No. 5147.
[9] A. N. vol. 206, No. 4945.

TABLE 20.—*Elements (884)—Priamus (1917 CQ.)*

Letter	Epoch	M.T.	M	ω	Ω	i
			° ′ ″	° ′ ″	° ′ ″	° ′ ″
A......	1917 Sept. 24.5.........	Greenwich..	83 18 55	329 32 38	300 41 27	8 51 26
B......	1917 Sept. 24.5.........	Greenwich..	83 18 55	329 32 18	300 48 28	8 51 28
C......	1917 Sept. 24.5.........	Greenwich..	83 46 41	329 4 44	300 48 28	8 51 28
D......	1917 Sept. 24.5.........	Greenwich..	82 22 47	329 51 19	300 49 27	8 51 24
E......	1918 Oct. 29.5.........	Greenwich..	115 11 59	329 45 49	300 49 27	8 51 24
F......	1918 Oct. 29.5.........	Greenwich..	115 33 19	329 24 42	300 49 27	8 51 24

Letter	Epoch	M.T.	φ	μ	Equinox	Authority
			° ′ ″	″		
A......	1917 Sept. 24.5.......	Greenwich..	6 46 53	294.427	1917.0	Berberich
B......	1917 Sept. 24.5.......	Greenwich..	6 46 53	294.427	1925.0	Berberich
C......	1917 Sept. 24.5.......	Greenwich..	6 46 54	294.989	1925.0	Berberich-Klose
D......	1917 Sept. 24.5.... ..	Greenwich..	7 5 53	294.427	1925.0	Klose
E......	1918 Oct. 29.5.......	Greenwich..	7 5 59	294 5850	1925.0	Klose
F......	1918 Oct. 29.5.......	Greenwich..	7 7 34	295.0965	1925.0	Klose

TABLE 21.—*References for Observations of (884) Priamus (1917 CQ.)*

	Date	Place	Reference
1917...	September 22...	Heidelberg....	A.N. 205 p. 141, M.N. 78 p. 289* E.Z. 1917 No. 535
	September 23...	Heidelberg....	A.N. 205 p. 141; E.Z. 1917 No. 535
	September 24...	Heidelberg....	A.N. 205 p. 141; E.Z. 1917 No. 535
	September 25...	Heidelberg....	A.N. 205 p. 141; E.Z. 1917 No. 535
	September 26...	Heidelberg....	A.N. 205 p. 141; E.Z. 1917 No. 535
	October 6...	Wien........	A.N. 207 p. 149
	October 11...	Wien........	A.N. 207 p. 149
	October 13...	Wien........	A.N. 207 p. 149
	October 16...	Wien........	A.N. 207 p. 149
	October 21...	Heidelberg....	A.N. 205 p. 239 E.Z. 1917 No. 537
	November 8...	Heidelberg....	A.N. 205 p. 239 E.Z. 1918 No. 537 A.N. 206 p. 63
	December 4...	Heidelberg....	A.N. 205 p. 279 E.Z. 1918 No. 538 A.N. 206 p. 63
1917...	December 4...	Bergedorf....	A.N. 208 p. 39 E.Z. 1918 No. 559
1918...	January 2...	Bergedorf....	A N. 208 p. 39 E.Z. 1918 No. 559
	January 3...	Heidelberg....	A.N. 206 p. 63 A.N. 206 p. 15
	January 14...	Bergedorf....	A.N. 208 p. 39 E.Z. 1918 No. 559
	October 5...	Heidelberg....	A.N. 207 p. 239 E.Z. 1918 No. 555
	October 30...	Heidelberg....	A.N. 207 p. 283 E.Z. 1918 No. 557
	November 23...	Heidelberg....	A.N. 208 p. 13 E.Z. 1918 No. 558 A.N. 208 p. 167
1919...	October 21...	Heidelberg....	B.Z. 1919 No. 13 Vol. 1
1921...	January 15...	Heidelberg....	B.Z. 1921 No. 3 Vol. 3

*Discovery date.

(911) AGAMEMNON, 1919 FD.

This planet was discovered by Reinmuth[1] at Heidelberg on March 19, 1919.

The nature of its motion (Jupiter group) was noted about the same time by Palisa and Berberich.[2] The preliminary elements A by Berberich are published in A. N. Vol. 208, p. 332. A comparison between computation and observation[3] for 1919 May 20, gives $\Delta a = -4^s \, \Delta\delta = -2'$.

Elements A are brought forward to mean equinox of 1925.0 and with an ephemeris are published in Kleine Planeten for 1920, p. 25 and p. 47.

An observation[4] on March 11, 1920, gives a correction to the ephemeris of $+1^m.5$ in right ascension and $-18'$ in declination.

An ephemeris for 1921 is given in Kleine Planeten for 1921, p. 20.

REFERENCES

[1] A. N. vol. 208, p. 231. Royal A. S. 210, p. 241.
vol. 80, p. 405, 406.   [3] A. N. vol. 210, p. 244.
[2] A. N. vol. 208, p. 331. A. N. vol.   [4] B-Z der A. N. No. 13, 1920.

Elements—(911) Agamemnon (1919 FD.)

	Epoch	M.T.	M.	ω	Ω	i
A	1919 Mar. 19.5	Grw.	88°48'19"	78°46'08"	336°55'10"	21°56'50"

φ	μ	Equinox	Authority	Remarks
4°55'43"	303".190	1919.0	Berberich.	Preliminary orbit.

TABLE 22.—*References for Observations of (911) Agamemnon (1919 FD)*

	Date	Place	Reference
1919...	March 19......	Königstuhl...	A.N. 208 p. 231*
	April 2......	Königstuhl...	A.N. 208 p. 231
	May 20......	Königstuhl...	A.N. 208 p. 347
	April 5......	Wien........	A.N. 211 p. 430
	April 6......	Wien........	A.N. 211 p. 430
	April 19......	Wien........	A.N. 211 p. 430
	April 19......	Wien........	A.N. 211 p. 430
	April 23......	Wien........	A.N. 211 p. 430
	April 25......	Wien........	A.N. 211 p. 430
	April 29......	Wien........	A.N. 211 p. 430
	May 1......	Wien........	A.N. 211 p. 430
1920...	March 11......	Bergedorf....	B. Z. 1920, No. 13

*Discovery Date.

www.ingramcontent.com/pod-product-compliance
Lightning Source LLC
Chambersburg PA
CBHW032016190326
41520CB00007B/495